RabbitMQ Cookbook

Over 70 practical recipes to help you develop messaging
applications using RabbitMQ with the help of plenty of
real-life examples

Sigismondo Boschi

Gabriele Santomaggio

BIRMINGHAM - MUMBAI

RabbitMQ Cookbook

First published: December 2013

Production Reference: 1171213

Published by Packt Publishing Ltd.

Livery Place
35 Livery Street
Birmingham B3 2PB, UK.

ISBN 978-1-84951-650-1

www.packtpub.com

Cover Image by Sandeep Vaity (sandeep.vaity@yahoo.com)

Credits

Authors
Sigismondo Boschi

Gabriele Santomaggio

Reviewers
Lucas Hrabovsky

Jorge Puente Sarrín

Ken Taylor

Ignacio Colomina Torregrosa

Héctor Veiga

Acquisition Editors
Usha Iyer

Llewellyn Rozario

Lead Technical Editor
Chalini Snega Victor

Copy Editors
Alisha Aranha

Roshni Banerjee

Dipti Kapadia

Karuna Narayanan

Kirti Pai

Shambhavi Pai

Alfida Paiva

Laxmi Subramanian

Technical Editors
Shubhangi Dhamgaye

Novina Kewalramani

Pratik More

Suneeth Nair

Rohit Kumar Singh

Project Coordinator
Anugya Khurana

Proofreader
Linda Morris

Indexer
Tejal Soni

Graphics
Ronak Dhruv

Abhinash Sahu

Production Coordinator
Adonia Jones

Cover Work
Adonia Jones

About the Authors

Sigismondo Boschi is a software developer currently involved in projects of messaging and networking distributed applications.

Prior to this, he has had more than 10 years' experience working with distributed applications and message-passing paradigms. He first acquired a PhD in Computational Physical Chemistry from the University of Bologna, and then has worked in the Development of Scientific High Performance Computing Projects.

To my wife, Maria Luisa, and my children, Gaia and Marco.

Gabriele Santomaggio has worked in the IT industry for more than 15 years. He is a developer, and very keen on middleware and distributed applications. Currently, he is working on high-performance Java applications.

He is the member of a big IT Italian community (www.indigenidigitali.com/) where he has published some posts about Amazon Web Services and message-oriented middleware (http://blog.indigenidigitali.com/tag/gabriele-santomaggio/).

He likes running and listening to jazz music.

To my beautiful wife, Dia, and my little ninja turtle, Riccardo, with love.

About the Reviewers

Jorge Puente Sarrín is a Peruvian software developer, currently working at Red Científica Peruana (RCP) as a software architect designing service-based solutions. He is a passionate developer focused on building distributed systems solutions using asynchronous programming with Python and .NET. Also, he has been contributing towards the translation of documentation projects and online courses into Spanish. He is a proud member of *Masters of MongoDB*, a group of persons who are promoting MongoDB around the world.

Ken Taylor has worked in software development and technology for over 15 years. During the course of his career, he has worked as a systems analyst on multiple software projects in several industries as well as with the state government. He has used RabbitMQ for messaging on multiple projects as a way of scaling and as an integration point. He is a member and speaker of the 757 Ruby users group and the Hampton Roads .NET users group (HRNUG). Ken holds an A.S. in Computer Science from Paul D. Camp Community College and was awarded a U.S. Patent for a Real Estate financial software product. He is currently working at Outsite Networks Inc. in Norfolk, Virginia. He lives in Suffolk, Virginia with his lovely wife Lucia and his two sons, Kaide and Wyatt.

I would like to thank my wife for her support while writing this book and my sons for reminding me of the importance of being inquisitive. Thank you Packt Publishing for asking me to participate as a technical reviewer for this excellent resource on RabbitMQ.

Ignacio Colomina Torregrosa is a technical engineer in telecommunications and a master in free software. He works as a PHP/Symfony developer and has had experience using RabbitMQ as a tool to optimize and improve the performance of web applications that deal with large amounts of traffic.

Héctor Veiga is a software engineer specializing in real-time data integration. Recently, he has focused his work on different cloud technologies (AWS, Heroku, OpenShift, and so on) to develop scalable, resilient, and high-performing applications that are able to handle high-volume, real-time data in diverse protocols and formats. Additionally, he has a strong knowledge of messaging systems, such as RabbitMQ and AMQP. Also, Héctor has a Master's degree in Telecommunication Engineering from the Universidad Politécnica de Madrid and a Master's degree in Information Technology and Management from the Illinois Institute of Technology.

Héctor currently works at HERE as part of Global Data Integration team and is actively developing scalable applications to consume data from several different sources. HERE heavily utilizes RabbitMQ to address their messaging requirements. In the past, Héctor has worked at Xaptum Technologies, a company dedicated to M2M technologies.

I would like to thank my family and friends for their support. They showed me that giving up was not an option and that you just need to keep pushing. Also, I am greatly thankful for the opportunity I was given to live and work in a foreign country. Finally, I would like to acknowledge Steve Fosdal who has been my friend and RabbitMQ colleague during this journey.

www.PacktPub.com

Support files, eBooks, discount offers and more

You might want to visit www.PacktPub.com for support files and downloads related to your book.

Did you know that Packt offers eBook versions of every book published, with PDF and ePub files available? You can upgrade to the eBook version at www.PacktPub.com and as a print book customer, you are entitled to a discount on the eBook copy. Get in touch with us at service@packtpub.com for more details.

At www.PacktPub.com, you can also read a collection of free technical articles, sign up for a range of free newsletters and receive exclusive discounts and offers on Packt books and eBooks.

http://PacktLib.PacktPub.com

Do you need instant solutions to your IT questions? PacktLib is Packt's online digital book library. Here, you can access, read and search across Packt's entire library of books.

Why Subscribe?

- ▶ Fully searchable across every book published by Packt
- ▶ Copy and paste, print and bookmark content
- ▶ On demand and accessible via web browser

Free Access for Packt account holders

If you have an account with Packt at www.PacktPub.com, you can use this to access PacktLib today and view nine entirely free books. Simply use your login credentials for immediate access.

Table of Contents

Preface

The RabbitMQ Cookbook covers RabbitMQ and the corresponding AMQP protocol.

Even the less-experienced programmer can find a lot of information that is useful to start developing messaging applications using RabbitMQ, especially in the first few chapters.

However, in order to fully appreciate some of the chapters of the book, the reader should also know the basic concepts about web and cloud applications.

The book is full of examples that show you how to use RabbitMQ in very disparate scenarios, in different technologies, right from single hosts to geographically replicated high-availability clusters. Mostly everyone working with these technologies will find a lot of useful information within this book.

What this book covers

Chapter 1, *Working with AMQP*, introduces the basic concepts of AMQP, the Advanced Message Queuing Protocol, on which RabbitMQ is based. It shows how the fire-and-forget messaging model works and how to use it from clients.

Chapter 2, *Going beyond the AMQP Standard*, covers the RabbitMQ extensions of AMQP and how they can be used to increase the efficiency of messaging applications.

Chapter 3, *Managing RabbitMQ*, explains how to configure RabbitMQ parameters, enable RabbitMQ plugins, and monitor RabbitMQ activities.

Chapter 4, *Mixing Different Technologies*, covers the integration of different technologies with other protocols, such as MQTT, STOMP, and JSON.

Chapter 5, *Using RabbitMQ in Web Applications*, shows how to develop web applications with RabbitMQ as a service bus.

Chapter 6, *Developing Scalable Applications*, explains how to create scalable and robust applications using RabbitMQ clustering.

Chapter 7, Developing High-availability Applications, covers the high-availability options offered by RabbitMQ.

Chapter 8, Performance Tuning for RabbitMQ, explains how to optimize the performance of applications based on RabbitMQ in different scenarios.

Chapter 9, Extending RabbitMQ Functionality, covers the use and development of RabbitMQ plugins.

Chapter 10, RabbitMQ on AWS, explains how to deploy RabbitMQ applications on Amazon Web Services.

Chapter 11, AMQP and Cloud Computing – RabbitMQ on PaaS, explains how to deploy and use RabbitMQ on Cloud Foundry.

Chapter 12, Managing RabbitMQ Error Conditions, explains how to diagnose and deal with RabbitMQ error conditions.

What you need for this book

You can use Linux/Unix, Mac OS X, or Windows as the operating system since RabbitMQ is cross-platform middleware. Most of the examples have been developed in Java, but to execute other recipes, you will also need:

  ▶ Python
  ▶ Ruby
  ▶ .NET (here you will be bound to Windows only)
  ▶ Erlang
  ▶ Objective-C
  ▶ JavaScript

The RabbitMQ Cookbook also covers the cloud computing world (both IaaS and PaaS). To utilize the respective chapters, you will need an account on:

  ▶ Amazon Web Services (IaaS)
  ▶ Cloud Foundry (PaaS)

Who this book is for

The RabbitMQ Cookbook is for software developers who want to develop distributed applications based on messaging. It's assumed that the reader has some experience with multithreaded applications and distributed applications.

The reader should also know the basic concepts of web and cloud applications in order to follow the recipes specifically.

Conventions

In this book, you will find a number of styles of text that distinguish between different kinds of information. Here are some examples of these styles, and an explanation of their meaning.

Code words in text, folder names, filenames, file extensions, pathnames, are shown as follows: "Properly configure CLASSPATH and your preferred development environment".

A block of code is set as follows:

```
ConnectionFactory factory = new ConnectionFactory();
String uri="amqp://user:pass@hostname:port/vhost";
factory.setUri(uri);
```

Any command-line input or output is written as follows:

```
java -cp ./bin  rmqexample.Publish [Rabbitmq-host]
```

New terms and **important words** are shown in bold. Words that you see on the screen, in menus or dialog boxes for example, appear in the text like this: "Configure the host by navigating to **Edit configuration | services | endpoint | new**."

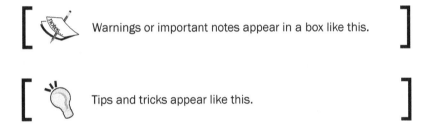

> Warnings or important notes appear in a box like this.

> Tips and tricks appear like this.

Reader feedback

Feedback from our readers is always welcome. Let us know what you think about this book— what you liked or may have disliked. Reader feedback is important for us to develop titles that you really get the most out of.

To send us general feedback, simply send an e-mail to feedback@packtpub.com, and mention the book title via the subject of your message.

If there is a topic that you have expertise in and you are interested in either writing or contributing to a book, see our author guide on www.packtpub.com/authors.

Customer support

Now that you are the proud owner of a Packt book, we have a number of things to help you to get the most from your purchase.

Downloading the example code

You can download the example code files for all Packt books you have purchased from your account at http://www.packtpub.com. If you purchased this book elsewhere, you can visit http://www.packtpub.com/support and register to have the files e-mailed directly to you.

Errata

Although we have taken every care to ensure the accuracy of our content, mistakes do happen. If you find a mistake in one of our books—maybe a mistake in the text or the code—we would be grateful if you would report this to us. By doing so, you can save other readers from frustration and help us improve subsequent versions of this book. If you find any errata, please report them by visiting http://www.packtpub.com/submit-errata, selecting your book, clicking on the **errata submission form** link, and entering the details of your errata. Once your errata are verified, your submission will be accepted and the errata will be uploaded on our website, or added to any list of existing errata, under the Errata section of that title. Any existing errata can be viewed by selecting your title from http://www.packtpub.com/support.

Piracy

Piracy of copyright material on the Internet is an ongoing problem across all media. At Packt, we take the protection of our copyright and licenses very seriously. If you come across any illegal copies of our works, in any form, on the Internet, please provide us with the location address or website name immediately so that we can pursue a remedy.

Please contact us at copyright@packtpub.com with a link to the suspected pirated material.

We appreciate your help in protecting our authors, and our ability to bring you valuable content.

Questions

You can contact us at questions@packtpub.com if you are having a problem with any aspect of the book, and we will do our best to address it.

Working with AMQP

1

In this chapter we will cover:

- ▶ Connecting to a broker
- ▶ Producing messages
- ▶ Consuming messages
- ▶ Using body serialization with JSON
- ▶ Using RPC with messaging
- ▶ Broadcasting messages
- ▶ Working with message routing using direct exchanges
- ▶ Working with message routing using topic exchanges
- ▶ Guaranteeing message processing
- ▶ Distributing messages to many consumers
- ▶ Using message properties
- ▶ Messaging with transactions
- ▶ Handling unroutable messages

Introduction

Advanced Message Queuing Protocol (**AMQP**) has been developed because of the need for interoperability among the many different messaging solutions, that were developed a few years ago by many different vendors such as IBM MQ-Series, TIBCO, or Microsoft MSMQ.

The AMQP 0-9-1 standard gives a complete specification of the protocol, particularly regarding:

- ▸ The API interface
- ▸ The wire protocol

RabbitMQ is a free and complete AMQP broker implementation. It implements version 0-9-1 of the AMQP specification; this is the most widespread version today and it is the last version that focuses on the client API. That's what we want to put the focus on, especially in this chapter.

On the other hand, AMQP 1.0 only defines the evolution of the wire-level protocol—the format of the data being passed at the application level—for the exchange of messages between two endpoints; so 0-9-1 is actually the most updated client library specification.

RabbitMQ includes:

- ▸ The broker itself, that is, the service that will actually handle the messages that are going to be sent and received by the applications
- ▸ The API implementations for Java, C#, and Erlang languages

It is also possible to use APIs for languages downloadable from the RabbitMQ site itself, from third-party sites, or even using AMQP APIs not strictly related to RabbitMQ (`http://www.rabbitmq.com/devtools.html`). Since the AMQP standard specifies the wire protocol, they are going to be mostly interoperable, except for some custom extensions. That will be discussed in detail in the next chapter.

In the course of the book we will particularly use some of the following APIs:

- ▸ The Java AMQP client library (`http://www.rabbitmq.com/java-client.html`)
- ▸ Pika, one of the Python AMQP client libraries (`http://pypi.python.org/pypi/pika`)
- ▸ The .NET/C# AMQP client (`http://www.rabbitmq.com/dotnet.html`)
- ▸ The RabbitMQ C client API (`https://github.com/alanxz/rabbitmq-c`)
- ▸ The Ruby client library (`https://github.com/ruby-amqp/bunny`)

In this first chapter we are mainly using Java since this language is widely used in enterprise software development, integration, and distribution. RabbitMQ is a perfect fit in this environment.

In order to run the examples in this recipe, you will first need to:

- ▸ Install Java JDK 1.6+
- ▸ Install the Java RabbitMQ client library

▶ Properly configure `CLASSPATH` and your preferred development environment (Eclipse, NetBeans, and so on)

▶ Install the RabbitMQ server on a machine (this can be the same local machine)

The natural choice is to install it on your desktop (Windows, Linux, and Mac OS X are all fine choices), but you can also install it on one or more external servers; for example, virtual machines, physical servers, and Raspberry PI servers (`http://www.raspberrypi.org/`) on cloud service providers.

In this book we are not providing instructions on the installation of RabbitMQ itself. You can find detailed instructions on the RabbitMQ site.

Most of the examples will work connecting to the RabbitMQ broker running on the localhost. If you have chosen to install or use RabbitMQ from a different machine, you will need to specify its hostname as a command-line parameter of the examples themselves, for example:

```
java -cp ./bin  rmqexample.Publish [Rabbitmq-host]
```

For the examples involving Python, you will need Python 2.7+ installed and the Pika library, an AMQP implementation for Python (`https://pypi.python.org/pypi/pika`). The fastest way to install Pika is by using PIP (`https://pypi.python.org/pypi/pip`). In the command prompt, just type:

```
pip install pika
```

We will also present some recipes using .NET where the accent is mainly on interoperability.

You can download the working examples in their full form at `http://www.packtpub.com/support`.

The recipes presented in this chapter will tackle all the basic concepts exposed by AMQP, using RabbitMQ.

Connecting to the broker

Every application that uses AMQP needs to establish a connection with the AMQP broker. By default, RabbitMQ (as well as any other AMQP broker up to version 1.0) works over TCP as a reliable transport protocol on port 5672, that is, the IANA-assigned port.

We are now going to discuss how to create the connection. In all the subsequent recipes we will refer to the connection and channel as the results of the operations presented here.

Getting ready

To use this recipe we need to set up the Java development environment as mentioned in the *Introduction* section.

How to do it...

In order to create a Java client that connects to the RabbitMQ broker, you need to perform the following steps:

1. Import the needed classes from the Java RabbitMQ client library in the program namespace:

    ```
    import com.rabbitmq.client.Channel;
    import com.rabbitmq.client.Connection;
    import com.rabbitmq.client.ConnectionFactory;
    ```

2. Create an instance of the client `ConnectionFactory`:

    ```
    ConnectionFactory factory = new ConnectionFactory();
    ```

3. Set the `ConnectionFactory` options:

    ```
    factory.setHost(rabbitMQhostname);
    ```

4. Connect to the RabbitMQ broker:

    ```
    Connection connection = factory.newConnection();
    ```

5. Create a channel from the freshly created connection:

    ```
    Channel channel = connection.createChannel();
    ```

6. As soon as we are done with RabbitMQ, release the channel and the connection:

    ```
    channel.close();
    connection.close();
    ```

How it works...

Using the Java client API, the application must create an instance of `ConnectionFactory` and set the host where RabbitMQ should be running with the `setHost()` method.

After the Java imports (step 1), we have instantiated the `factory` object (step 2). In this example we have just set the hostname that we have chosen to optionally get from the command line (step 3), but you can find more information regarding connection options in the section *There's more...*.

In step 4 we have actually established the TCP connection to the RabbitMQ broker.

> In this recipe we have used the default connection parameters user: `guest`, password: `guest`, and vhost: `/`; we will discuss these parameters later.

However, we are not yet ready to communicate with the broker; we need to set up a communication channel (step 5). This is an advanced concept of AMQP; using this abstraction, it is possible to let many different messaging sessions use the same logical connection.

Actually, all the communication operations of the Java client library are performed by the methods of a channel instance.

If you are developing multithreaded applications, it is highly recommended to use a different channel for each thread. If many threads use the same channel, they will serialize their execution in the channel method calls, leading to possible performance degradation.

> The best practice is to open a connection and share it with different threads. Each thread creates, uses, and destroys its own independent channel(s).

There's more...

It is possible to specify many different optional properties for any RabbitMQ connection. You can find them all in the online documentation at (http://www.rabbitmq.com/releases/rabbitmq-java-client/current-javadoc). These options are all self-explanatory, except for the AMQP virtual host.

Virtual hosts are administrative containers; they allow to configure many logically independent brokers hosts within one single RabbitMQ instance, to let many different independent applications share the same RabbitMQ server. Each virtual host can be configured with its independent set of permissions, exchanges, and queues and will work in a logically separated environment.

It's possible to specify connection options by using just a connection string, also called **connection URI**, with the `factory.setUri()` method:

```
ConnectionFactory factory = new ConnectionFactory();
String uri="amqp://user:pass@hostname:port/vhost";
factory.setUri(uri);
```

> The URI must conform to the syntax specified in RFC3986 (http://www.ietf.org/rfc/rfc3986.txt).

Producing messages

In this recipe we are learning how to send a message to an AMQP queue. We will be introduced to the building blocks of AMQP messaging: messages, queues, and exchanges.

You can find the source at `Chapter01/Recipe02/src/rmqexample`.

Getting ready

To use this recipe we need to set up the Java development environment as indicated in the *Introduction* section.

How to do it...

After connecting to the broker, as seen in the previous recipe, you can start sending messages performing the following steps:

1. Declare the queue, calling the `queueDeclare()` method on `com.rabbitmq.client.Channel`:

   ```
   String myQueue = "myFirstQueue";
   channel.queueDeclare(myQueue, true, false, false, null);
   ```

2. Send the very first message to the RabbitMQ broker:

   ```
   String message = "My message to myFirstQueue";
   channel.basicPublish("",myQueue, null, message.getBytes());
   ```

3. Send the second message with different options:

   ```
   channel.basicPublish("",myQueue,MessageProperties.
   PERSISTENT_TEXT_PLAIN,message.getBytes());
   ```

 The queue names are case sensitive: `MYFIRSTQUEUE` is different from `myFirstQueue`.

How it works...

In this first basic example we have been able to just send a message to RabbitMQ.

After the communication channel is established, the first step is to ensure that the destination queue exists. This task is accomplished declaring the queue (step 1) calling `queueDeclare()`. The method call does nothing if the queue already exists, otherwise it creates the queue itself.

If the queue already exists but has been created with different parameters, `queueDeclare()` will raise an exception.

Note that this, as most of the AMQP operations, is a method of the `Channel` Java interface. All the operations that need interactions with the broker are carried out through channels.

Let's examine the meaning of the `queueDeclare()` method call in depth. Its template can be found in the Java client reference documentation located at `http://www.rabbitmq.com/releases/rabbitmq-java-client/current-javadoc/`. The documentation will be as shown in the following screenshot:

In particular we have used the second overload of this method that we report here:

```
AMQP.Queue.DeclareOk queueDeclare(java.lang.String queue, boolean
    durable, boolean exclusive, booleanautoDelete,
     java.util.Map<java.lang.String,java.lang.Object> arguments)
       throws java.io.IOException
```

The meanings of the individual arguments are:

- `queue`: This is just the name of the queue where we will be storing the messages.
- `durable`: This specifies whether the queue will survive server restarts. Note that it is required for a queue to be declared as durable if you want persistent messages to survive a server restart.

- ▸ `exclusive`: This specifies whether the queue is restricted to only this connection.
- ▸ `autoDelete`: This specifies whether the queue will be automatically deleted by the RabbitMQ broker as soon as it is not in use.
- ▸ `arguments`: This is an optional map of queue construction arguments.

In step 2 we have actually sent a message to the RabbitMQ broker.

The message body will never be opened by RabbitMQ. Messages are opaque entities for the AMQP broker, and you can use any serialization format you like. We often use JSON, but XML, ASN.1, standard or custom, ASCII or binary format, are all valid alternatives. The only important thing is that the client applications should know how to interpret the data.

Let's now examine in depth the `basicPublish()` method of the `Channel` interface for the overload used in our recipe:

```
void basicPublish(java.lang.String exchange,
    java.lang.String routingKey, AMQP.BasicProperties props, byte[]
        body) throws java.io.IOException
```

In our example the `exchange` argument has been set to the empty string `""`, that is, the default exchange, and the `routingKey` argument to the name of the queue. In this case the message is directly sent to the queue specified as `routingKey`. The `body` argument is set to the `byte` array of our string, that is, just the message that we sent. The `props` argument is set to `null` as a default; these are the message properties, discussed in depth in the recipe *Using message properties*.

For example, in step 3 we have sent an identical message, but with `props` set to `MessageProperties.PERSISTENT_TEXT_PLAIN`; in this way we have requested RabbitMQ to mark this message as a **persistent message**.

Both the messages have been dispatched to the RabbitMQ broker, logically queued in the `myFirstQueue` queue. The messages will stay buffered there until a client, (typically, a different client) gets it.

If the queue has been declared with the `durable` flag set to `true` and the message has been marked as persistent, it is stored on the disk by the broker. If one of the two conditions is missing, the message is stored in the memory. In the latter case the buffered messages won't survive a RabbitMQ restart, but the message delivery and retrieval will be much faster. However, we will dig down on this topic in *Chapter 8, Performance Tuning for RabbitMQ*.

There's more...

In this section we will discuss the methods to check the status of RabbitMQ and whether a queue already exists.

How to check the RabbitMQ status

In order to check the RabbitMQ status, you can use the command-line control tool `rabbitmqctl`. It should be in the `PATH` in the Linux setup. On Windows it can be found running the RabbitMQ command shell by navigating to **Start Menu | All Programs | RabbitMQ Server | RabbitMQ Command Prompt (sbin dir)**. We can run `rabbitmqctl.bat` from this command prompt.

We can check the queue status with the command `rabbitmqclt list_queues`. In the following screenshot, we have run it just before and after we have run our example.

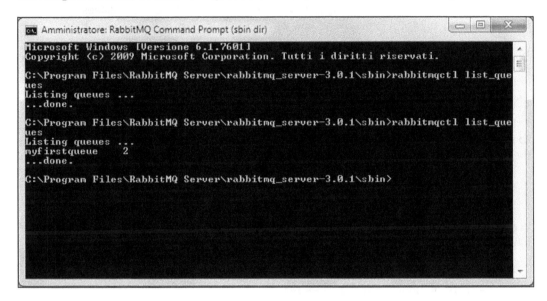

We can see our **myfirstqueue** queue listed in the preceding screenshot, followed by the number **2**, which is just the number of the messages buffered into our queue.

Now we can either try to restart RabbitMQ, or reboot the machine hosting it. Restarting RabbitMQ successfully will depend on the used OS:

- On Linux, RedHat, Centos, Fedora, Raspbian, and so on:

  `service rabbitmq-server restart`

- On Linux, Ubuntu, Debian, and so on:

  `/etc/init.d/rabbitmq restart`

- On Windows:

  `sc stop rabbitmq / sc start rabbitmq`

How many messages should we expect when we run `rabbitmqclt list_queues` again?

Checking whether a queue already exists

In order to be sure that a specific queue already exists, replace `channel.queueDeclare()` with `channel.queueDeclarePassive()`. The behavior of the two methods is the same in case the queue already exists; but in case it doesn't, the first one will silently create it and return back (that's actually the most frequently used case), the latter will raise an exception.

Consuming messages

In this recipe we are closing the loop; we have already seen how to send messages to RabbitMQ—or to any AMQP broker—and now we are ready to learn how to retrieve them.

You can find the source code of the recipe at `Chapter01/Recipe03/src/rmqexample/nonblocking`.

Getting ready

To use this recipe we need to set up the Java development environment as indicated in the introduction.

How to do it...

In order to consume the messages sent as seen in the previous recipe, perform the following steps:

1. Declare the queue where we want to consume the messages from:

   ```
   String myQueue="myFirstQueue";
   channel.queueDeclare(myQueue, true, false, false, null);
   ```

2. Define a specialized consumer class inherited from `DefaultConsumer`:

   ```
   public class ActualConsumer extends DefaultConsumer {
     public ActualConsumer(Channel channel) {
       super(channel);
     }
     @Override
     public void handleDelivery(
       String consumerTag,
       Envelope envelope,
       BasicProperties properties,
       byte[] body) throws java.io.IOException {
         String message = new String(body);
         System.out.println("Received: " + message);
       }
   }
   ```

3. Create a `consumer` object, which is an instance of this class, bound to our channel:

   ```
   ActualConsumer consumer = new ActualConsumer(channel);
   ```

4. Start consuming messages:

   ```
   String consumerTag = channel.basicConsume(myQueue, true,
       consumer);
   ```

5. Once done, stop the consumer:

   ```
   channel.basicCancel(consumerTag);
   ```

How it works...

After we have established the connection and the channel to the AMQP broker as seen in the *Connecting to the broker* recipe, we need to ensure that the queue from which we are going to consume the messages exists (step 1).

In fact it is possible that the consumer is started before any producer has sent a message to the queue and the queue itself may actually not exist at all. To avoid the failure of the subsequent operations on the queue, we need to declare the queue.

 By allowing both producers and consumers to declare the same queue, we are decoupling their existence; the order in which we start them is not important.

The heart of this recipe is step 2. Here we have defined our specialized consumer that overrides `handleDelivery()` and instantiated it in step 3. In the Java client API the consumer callbacks are defined by the `com.rabbitmq.client.Consumer` interface. We have extended our consumer from `DefaultConsumer`, which provides a no-operation implementation for all the methods declared in the `Consumer` interface.

In step 3, by calling `channel.basicConsume()`, we let the **consumer threads** start consuming messages. The consumers of each channel are always executed on the same thread, independent of the calling one.

Now that we have activated a consumer for `myQueue`, the Java client library starts getting messages from that queue on the RabbitMQ broker, and invokes `handleDelivery()` for each one.

Then after the `channel.basicConsume()` method's invocation, we just sit idle waiting for a key press in the main thread. Messages are being consumed with **nonblocking semantics** respecting the **event-driven** paradigm, typical of messaging applications.

Only after we press *Enter*, the execution proceeds to step 5, cancelling the consumer. At this point the consumer threads stop invoking our consumer object, and we can release the resources and exit.

There's more...

In this section we will learn more about consumer threads and the use of blockage semantics.

More on consumer threads

At connection definition time, the RabbitMQ Java API allocates a thread pool from which it will allocate consumer threads on need.

All the consumers bound to one channel will be executed by one single thread in the pool; however, it is possible that consumers from different channels are handled by the same thread. That's why it is important to avoid long-lasting operations in the consumer methods, in order to avoid blocking other consumers.

It is also possible to handle the consumer thread pool by ourselves, as we have shown in our example; however, this not obligatory at all. We have defined a thread pool, `java.util.concurrent.ExecutorService`, and passed it at connection time:

```
ExecutorService eService = Executors.newFixedThreadPool(10);
Connection connection = factory.newConnection(eService);
```

As we were managing it, we were also in charge of terminating it:

```
eService.shutdown();
```

However, remember that if you don't define your own `ExecutorService` thread pool, the Java client library will create one during connection creation time, and destroy it as soon as we destroy the corresponding connections.

Blocking semantics

It is possible to use blocking semantics too, but we strongly discourage this approach if it's not being used for simple applications and test cases; the recipe to consume messages is non-blocking.

However, you can find the source code for the blocking approach at `Chapter01/Recipe03/src/rmqexample/blocking`.

See also

You can find all the available methods of the consumer interface in the official Javadoc at

`http://www.rabbitmq.com/releases/rabbitmq-java-client/current-javadoc/com/rabbitmq/client/Consumer.html`

Using body serialization with JSON

In AMQP the messages are opaque entities; AMQP does not provide any standard way to encode/decode them.

However, web applications very often use JSON as an application layer format, that is, the JavaScript serialization format that has become a de-facto standard; in this way, the RabbitMQ client Java library can include some utility functions for this task.

On the other side, this is not the only protocol; any application can choose its own protocol (XML, Google Protocol Buffers, ASN.1, or proprietary).

In this example we are showing how to use the JSON protocol to encode and decode message bodies. We are using a publisher written in Java (`Chapter01/Recipe04/Java_4/src/rmqexample`) and a consumer in Python (`Chapter01/Recipe04/Python04`).

Getting ready

To use this recipe you will need to set up Java and Python environments as described in the introduction.

How to do it...

To implement a Java producer and a Python consumer, you can perform the following steps:

1. Java: In addition to the standard import described in the recipe *Connecting to the broker*, we have to import:

   ```
   import
   com.rabbitmq.tools.json.JSONWriter;
   ```

2. Java: We create a queue that is not persistent:

   ```
   String myQueue="myJSONBodyQueue_4";
     channel.queueDeclare(MyQueue, false, false, false, null);
   ```

3. Java: We create a list for the `Book` class and fill it with example data:

   ```
   List<Book>newBooks = new ArrayList<Book>();
     for (inti = 1; i< 11; i++) {
       Book book = new Book();
       book.setBookID(i);
       book.setBookDescription("History VOL: " + i  );
       book.setAuthor("John Doe");
       newBooks.add(book);
     }
   ```

4. Java: We are ready to serialize the `newBooks` instance with `JSONwriter`:

```
JSONWriter rabbitmqJson = new JSONWriter();
String jsonmessage = rabbitmqJson.write(newBooks);
```

5. Java: We can finally send our `jsonmessage`:

```
channel.basicPublish("",MyQueue,null,
    jsonmessage.getBytes());
```

6. Python: To use the Pika library we must add the follow import:

```
import pika;
import json;
```

Python has a powerful built-in library for JSON.

7. Python: In order to create a connection to RabbitMQ, use the following code:

```
connection =
    pika.BlockingConnection(pika.ConnectionParameters(rabbitmq_
        host));
```

8. Python: Let's declare a queue, bind as a consumer, and then register a callback:

```
channel = connection.channel()
my_queue = "myJSONBodyQueue_4"
channel.queue_declare(queue=my_queue)
channel.basic_consume(consumer_callback, queue=my_queue,
    no_ack=True)
channel.start_consuming()
```

How it works...

After we set up the environments (step 1 and step 2), we serialize the `newbooks` class with the method `write(newbooks)`. The method returns a JSON `String` (`jsonmessage`) as shown in the following code snippet:

```
[
  {
    "author" : "John Doe",
    "bookDescription" : "History VOL: 1",
    "bookID" : 1
  },
  {
    "author" : "John Doe",
    "bookDescription" : "History VOL: 2",
    "bookID" : 2
  }
]
```

In step 4 we publish jsonmessage to the queue myJSONBodyQueue_4. Now the Python Consumer can get the message from the same queue. Let's see how to do it in Python:

```
connection =
pika.BlockingConnection(pika.ConnectionParameters(rabbitmq_host));
channel = connection.channel()
queue_name = "myJSONBodyQueue_4"
channel.queue_declare(queue=my_queue)
..
channel.basic_consume(consumer_callback, queue=my_queue,
    no_ack=True)
channel.start_consuming()
```

As we have seen in the Java implementation, we must create a connection and then create a channel. With the method channel.queue_declare(queue=myQueue), we declare a queue that is not durable, exclusive or autodelete. In order to change the queue's attribute, it's enough to add the parameter in the queue_declare method as follows:

```
channel.queue_declare(queue=myQueue,durable=True)
```

 When different AMQP clients declare the same queue, it's important that all of them specify the same durable, exclusive, and autodelete attributes. Otherwise, channel.queue_declare() will raise an exception.

With the method channel.basic_consume(), the client starts consuming messages from the given queue, invoking the callback consumer_callback() where it will receive the messages.

While the callbacks in Java were defined in the consumer interface, in Python they are just passed to basic_consume(), in spite of the more functional, less declarative, and less formal paradigm typical of Python.

The callback consumer_callback is as follows:

```
def consumer_callback(ch, method, properties, body):
  newBooks=json.loads(body);
  print" Count books:",len(newBooks);
  for item in newBooks:
    print 'ID:',item['bookID'], '-
      Description:',item['bookDescription'],' -
        Author:',item['author']
```

The callback takes the message, deserializes it with json.loads(), and then the newBooks structure is ready to be read.

There's more...

The JSON helper tools included in the RabbitMQ client library are very simple, but in a real project you can evaluate them to use an external JSON library. For example, a powerful Java JSON library is google-gson (`https://code.google.com/p/google-gson/`) or jackson (`http://jackson.codehaus.org/`).

Using RPC with messaging

Remote Procedure Calls (**RPC**) are commonly used with client-server architectures. The client is required to perform some actions to the server, and then waits for the server reply.

The messaging paradigm tries to enforce a totally different approach with the fire-and-forget messaging style, but it is possible to use properly designed AMQP queues to perform and enhance RPC, as shown in the following figure:

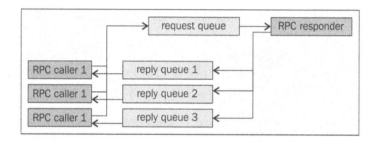

Graphically it is depicted that the request queue is associated with the responder, the reply queues with the callers.

However, when we use RabbitMQ, all the involved peers (both the callers and the responders) are AMQP clients.

We are now going to describe the steps performed in the example in `Chapter01/Recipe05/Java_5/src/rmqexample/rpc`.

Getting ready

To use this recipe we need to set up the Java development environment as indicated in the *Introduction* section.

How to do it...

Let's perform the following steps to implement the RPC responder:

1. Declare the `request queue` where the responder will be waiting for the RPC requests:

   ```
   channel.queueDeclare(requestQueue, false, false, false,
       null);
   ```

2. Define our specialized consumer `RpcResponderConsumer` by overriding `DefaultConsumer.handleDelivery()` as already seen in the *Consuming messages* recipe. On the reception of each RPC request, this consumer will:

   ❑ Perform the action required in the RPC request

   ❑ Prepare the `reply` message

   ❑ Set the correlation ID in the reply properties by using the following code:

   ```
   BasicProperties replyProperties = new
       BasicProperties.Builder().correlationId(properties.
       getCorrelationId()).build();
   ```

   ❑ Publish the answer on the `reply` queue:

   ```
   getChannel().basicPublish("", properties.getReplyTo(),
       replyProperties, reply.getBytes());
   ```

   ❑ Send the `ack` to the RPC `request`:

   ```
   getChannel().basicAck(envelope.getDeliveryTag(), false);
   ```

3. Start consuming messages, until we stop it as already seen in the *Consuming messages* recipe.

Now let's perform the following steps to implement the RPC caller:

1. Declare the request queue where the responder will be waiting for the RPC requests:

   ```
   channel.queueDeclare(requestQueue, false, false, false,
       null);
   ```

2. Create a temporary, private, autodelete reply queue:

   ```
   String replyQueue = channel.queueDeclare().getQueue();
   ```

3. Define our specialized consumer `RpcCallerConsumer`, which will take care of receiving and handling RPC replies. It will:

   ❑ Allow to specify what to do when it gets the replies (in our example, by defining `AddAction()`)

   ❑ Override `handleDelivery()`:

```
public void handleDelivery(String consumerTag,
Envelope envelope,
AMQP.BasicProperties properties,
byte[] body) throws java.io.IOException {

    String messageIdentifier =
      properties.getCorrelationId();
    String action = actions.get(messageIdentifier);
    actions.remove(messageIdentifier);

    String response = new String(body);
    OnReply(action, response);
}
```

4. Start consuming reply messages invoking `channel.basicConsume()`.

5. Prepare and serialize the requests (`messageRequest` in our example).

6. Initialize an arbitrary, unique message identifier (`messageIdentifier`).

7. Define what to do when the consumer gets the corresponding reply, by binding the action with the `messageIdentifier`. In our example we do it by calling our custom method `consumer.AddAction()`.

8. Publish the message to `requestqueue`, setting its properties:

```
BasicProperties props = new BasicProperties.Builder()
.correlationId(messageIdentifier)
.replyTo(replyQueue).build();
channel.basicPublish("", requestQueue,
  props,messageRequest.getBytes());
```

How it works...

In this example the RPC responder takes the role of an RPC server; the responder listens on the `requestQueue` public queue (step 1), where the callers will place their requests.

Each caller, on the other hand, will consume the responder replies on its own private queue, created in step 5.

When the caller sends a message (step 11), it includes two properties: the name of the temporary reply queue (`replyTo()`) where it will be listening, and a message identifier (`correlationId()`), needed by the caller to identify the call when the reply comes back.

In fact, in our example we have implemented an asynchronous RPC caller. The action to be performed by the `RpcCallerConsumer` (step 6) when the reply comes back is recorded by the nonblocking consumer by calling `AddAction()` (step 10).

Coming back to the responder, the RPC logic is all in the `RpcResponderConsumer`. This is not different from a specialized non-blocking consumer, as we have seen in the *Consuming messages* recipe, except for two details:

▸ The reply queue name is got by the message properties, `properties.getReplyTo()`. Its value has been set by the caller to its private, temporary reply queue.

▸ The reply message must include in its properties the correlation ID sent in the incoming message.

> The correlation ID is not used by the RPC responder; it is only used to let the caller receiving this message correlate this reply with its corresponding request.

There's more...

In this section we will discuss the use of blocking RPC and some scalability notes.

Using blocking RPC

Sometimes simplicity is more important than scalability. In this case it is possible to use the following helper classes, included in the Java RabbitMQ client library, that implement blocking RPC semantics:

```
com.rabbitmq.client.RpcClient
com.rabbitmq.client.StringRpcServer
```

The logic is identical, but there are no non-blocking consumers involved, and the handling of temporary queues and correlation IDs is transparent to the user.

You can find a working example at `Chapter01/Recipe05/Java_5/src/rmqexample/simplerpc`.

Scalability notes

What happens when there are multiple callers? It mainly works as a standard RPC client/server architecture. But what if we run many responders?

In this case all the responders will take care of consuming messages from the request queue. Furthermore, the responders can be located on different hosts. We have just got load distribution for free. More on this topic is in the recipe *Distributing messages to many consumers*.

Broadcasting messages

In this example we are seeing how to send the same message to a possibly large number of consumers.

This is a typical messaging application, broadcasting to a huge number of clients. For example, when updating the scoreboard in a massive multiplayer game, or when publishing news in a social network application.

In this recipe we are discussing both the producer and consumer implementation.

Since it is very typical to have consumers using different technologies and programming languages, we are using Java, Python, and Ruby to show interoperability with AMQP.

We are going to appreciate the benefits of having separated exchanges and queues in AMQP.

You can find the source in `Chapter01/Recipe06/`.

Getting ready

To use this recipe you will need to set up Java, Python and Ruby environments as described in the *Introduction* section.

How to do it...

To cook this recipe we are preparing four different codes:

- ▸ The Java publisher
- ▸ The Java consumer
- ▸ The Python consumer
- ▸ The Ruby consumer

To prepare a Java publisher:

1. Declare a `fanout` exchange:

    ```
    channel.exchangeDeclare(myExchange, "fanout");
    ```

2. Send one message to the exchange:

    ```
    channel.basicPublish(myExchange, "", null,
      jsonmessage.getBytes());
    ```

Then to prepare a Java consumer:

1. Declare the same `fanout` exchange declared by the producer:

   ```
   channel.exchangeDeclare(myExchange, "fanout");
   ```

2. Autocreate a new temporary queue:

   ```
   String queueName = channel.queueDeclare().getQueue();
   ```

3. Bind the queue to the exchange:

   ```
   channel.queueBind(queueName, myExchange, "");
   ```

4. Define a custom, non-blocking consumer, as already seen in the *Consuming messages* recipe.

5. Consume messages invoking `channel.basicConsume()`

The source code of the Python consumer is very similar to the Java consumer, so there is no need to repeat the needed steps. Just follow the steps of the Java consumer, looking to the source code in the archive of the recipes at:

`Chapter01/Recipe06/Python_6/PyConsumer.py`

In the Ruby consumer you need to use `require` "bunny" and then use the URI connection. Check out the source code at:

`Chapter01/Recipe06/Ruby_6/RbConsumer.rb`

We are now ready to mix all together, to see the recipe in action:

1. Start one instance of the Java producer; messages start getting published immediately.

2. Start one or more instances of the Java/Python/Ruby consumer; the consumers receive only the messages sent while they are running.

3. Stop one of the consumers while the producer is running, and then restart it; we can see that the consumer has lost the messages sent while it was down.

How it works...

Both the producer and the consumers are connected to RabbitMQ with a single connection, but the logical path of the messages is depicted in the following figure:

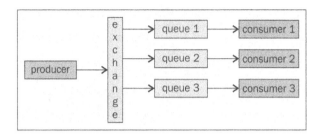

In step 1 we have declared the exchange that we are using. The logic is the same as in the queue declaration: if the specified exchange doesn't exist, create it; otherwise, do nothing.

The second argument of `exchangeDeclare()` is a string, specifying the type of the exchange, `fanout` in this case.

In step 2 the producer sends one message to the exchange. You can just view it along with the other defined exchanges issuing the following command on the RabbitMQ command shell:

`rabbitmqctl list_exchanges`

The second argument in the call to `channel.basicPublish()` is the **routing key**, which is always ignored when used with a `fanout` exchange. The third argument, set to `null`, is the optional message property (more on this in the *Using message properties* recipe). The fourth argument is just the message itself.

When we started one consumer, it created its own temporary queue (step 9). Using the `channel.queueDeclare()` empty overload, we are creating a nondurable, exclusive, autodelete queue with an autogenerated name.

Launching a couple of consumers and issuing `rabbitmqctl list_queues`, we can see two queues, one per consumer, with their odd names, along with the persistent `myFirstQueue` used in previous recipes as shown in the following screenshot:

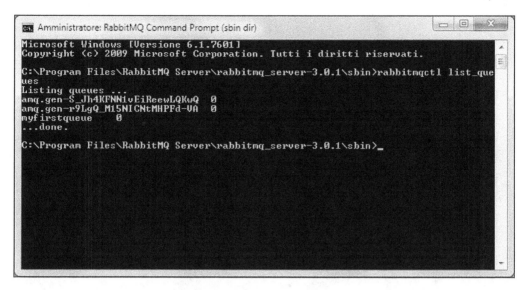

In step 5 we have bound the queues to `myExchange`. It is possible to monitor these **bindings** too, issuing the following command:

rabbitmqctl list_bindings

The monitoring is a very important aspect of AMQP; messages are routed by exchanges to the bound queues, and buffered in the queues.

 Exchanges do not buffer messages; they are just logical elements.

The `fanout` exchange routes messages by just placing a copy of them in each bound queue. So, no bound queues and all the messages are just received by no one consumer (see the *Handling unroutable messages* recipe for more details).

As soon as we close one consumer, we implicitly destroy its private temporary queue (that's why the queues are autodelete; otherwise, these queues would be left behind unused, and the number of queues on the broker would increase indefinitely), and messages are not buffered to it anymore.

When we restart the consumer, it will create a new, independent queue and as soon as we bind it to `myExchange`, messages sent by the publisher will be buffered into this queue and pulled by the consumer itself.

There's more...

When RabbitMQ is started for the first time, it creates some **predefined exchanges**. Issuing `rabbitmqctl list_exchanges` we can observe many existing exchanges, in addition to the one that we have defined in this recipe:

```
Amministratore RabbitMQ Command Prompt (sbin dir)

C:\Program Files\RabbitMQ Server\rabbitmq_server-3.0.1\sbin>rabbitmqctl list_exchanges
Listing exchanges ...
            direct
amq.direct        direct
amq.fanout        fanout
amq.headers       headers
amq.match         headers
amq.rabbitmq.log         topic
amq.rabbitmq.trace       topic
amq.topic         topic
lastnews.fanout   fanout
...done.

C:\Program Files\RabbitMQ Server\rabbitmq_server-3.0.1\sbin>
```

All the `amq.*` exchanges listed here are already defined by all the AMQP-compliant brokers and can be used instead of defining your own exchanges; they do not need to be declared at all.

We could have used `amq.fanout` in place of `myLastnews.fanout_6`, and this is a good choice for very simple applications. However, applications generally declare and use their own exchanges.

See also

With the overload used in the recipe, the exchange is non-autodelete (won't be deleted as soon as the last client detaches it) and non-durable (won't survive server restarts). You can find more available options and overloads at `http://www.rabbitmq.com/releases/rabbitmq-java-client/current-javadoc/`.

Working with message routing using direct exchanges

In this recipe we are going to see how to select a subset of messages to be consumed, routing them only to the AMQP queues of interest and ignoring all the others.

A typical use case is implementing a chat, where each queue represents a user.

We can find the relative example in the book examples directory at:

```
Chapter01/Recipe07/Java_7/src/rmqexample/direct
```

Getting ready

To use this recipe we need to set up the Java development environment as indicated in the *Introduction* section.

How to do it...

We are going to show how to implement both the producer and the consumer, and see them in action. To implement the producer, perform the following steps:

1. Declare a direct exchange:

    ```
    channel.exchangeDeclare(exchangeName, "direct", false,
        false, null);
    ```

2. Send some messages to the exchange, using arbitrary `routingKey` values:

    ```
    channel.basicPublish(exchangeName, routingKey, null,
        jsonBook.getBytes());
    ```

To implement the consumer, perform the following steps:

1. Declare the same exchange, identical to what was done in step 1.
2. Create a temporary queue:

    ```
    String myQueue = channel.queueDeclare().getQueue();
    ```

3. Bind the queue to the exchange using the `bindingKey`. Perform this operation as many times as needed, in case you want to use more than one binding key:

    ```
    channel.queueBind(myQueue,exchangeName,bindingKey);
    ```

4. After having created a suitable consumer object, start consuming messages as already seen in the *Consuming messages* recipe.

How it works...

In this recipe we have published messages (step 2) tagged with an arbitrary string (the so called routing key), to a direct exchange.

As with `fanout` exchanges, messages are not stored if there are no queues bound; however, in this case the consumers can choose the messages to be forwarded to these queues, depending on the binding key specified when they are bound (step 5).

Only messages with a routing key equal to the one specified in the binding will be delivered to such queues.

 This filtering operation is performed by the AMQP broker, and not by the consumer; the messages with a routing key that is different from the queue binding key won't be placed in that queue at all.

However, it's possible to have more queues bound with the same binding key; in this case, the broker will place a copy of the matching messages in all of them.

It is also possible to bind many different binding keys to the same queue/exchange pair, letting all the corresponding messages be delivered.

There's more...

In case we deliver a message with a given routing key to an exchange, and there are no queues bound with that specific key, the message is silently dropped.

However, the producer can detect and behave consequently when this happens, as shown in detail in the *Handling unroutable messages* recipe.

Working with message routing using topic exchanges

Direct and topic exchanges are conceptually very similar to each other. The main difference is that direct exchanges use **exact matching** only to select the destination of the messages, while topic exchanges allow using **pattern matching** with specific **wildcards**.

For example, the BBC is using topic routing with RabbitMQ to route new stories to all of the appropriate RSS feeds on their websites.

You can find the example for a topic exchange at:

```
Chapter01/Recipe08/Java_8/src/rmqexample/topic
```

Getting ready

To use this recipe we need to set up the Java development environment as indicated in the *Introduction* section.

How to do it...

Let's start with the producer:

1. Declare a topic exchange:

```
channel.exchangeDeclare(exchangeName, "topic", false,
    false, null);
```

2. Send some messages to the exchange, using arbitrary `routingKey` values:

```
channel.basicPublish(exchangeName, routingKey, null,
    jsonBook.getBytes());
```

Then, the consumers:

1. Declare the same exchange, identical to what was done in step 1.
2. Create a temporary queue:

```
String myQueue = channel.queueDeclare().getQueue();
```

3. Bind the queue to the exchange using the binding key, which in this case can contain wildcards:

```
channel.queueBind(myQueue,exchangeName,bindingKey);
```

4. After having created a suitable consumer object, start consuming messages as already seen in the *Consuming messages* recipe.

How it works...

As in the previous recipe, messages sent to a topic exchange are tagged with a string (step 2), but it is important for a topic exchange to be composed more of dot-separated words; these are supposed to be the topics of the message. For example, in our code we have used:

```
technology.rabbitmq.ebook
sport.golf.paper
sport.tennis.ebook
```

To consume these messages the consumer has to bind `myQueue` to the exchange (step 5) using the appropriate key.

 Using the messaging jargon, the consumer has to **subscribe** to the topics it's interested in.

Using the topic exchange, the subscription/binding key specified in step 5 can be a sequence of dot-separated words and/or wildcards. AMQP wildcards are just:

- ▶ `#`: This matches zero or more words
- ▶ `*`: This matches exactly one word

So, for example:

- ▶ `#.ebook` and `*.*.ebook` both match the first and the third sent messages
- ▶ `sport.#` and `sport.*.*` both match the second and the third sent messages
- ▶ `# alone` matches any message sent

In the last case the topic exchange behaves exactly like a `fanout` exchange, except for the performance, which is inevitably higher when using the former.

There's more...

Again, if some messages cannot be delivered to any one queue, they are silently dropped.

The producer can detect and behave consequently when this happens, as shown in detail in the *Handling unroutable messages* recipe.

Guaranteeing message processing

In this example we will show how to use the explicit acknowledgment, the so-called `ack`, while consuming messages.

A message is stored in a queue until one consumer gets the message and sends the `ack` back to the broker.

The `ack` can be either implicit or explicit. In the previous examples we have used the implicit `ack`.

In order to view this example in action, you can run the publisher from the *Producing messages* recipe and the consumer who gets the message, which you can find in the book archive at `Chapter01/Recipe09/Java_9/`.

Getting ready

To use this recipe we need to set up the Java development environment as indicated in the *Introduction* section.

How to do it...

In order to guarantee that the messages have been acknowledged by the consumer after processing them, you can perform the following steps:

1. Declare a queue:

```
channel.queueDeclare(myQueue, true, false, false,null);
```

2. Bind the consumer to the queue, specifying `false` for the `autoAck` parameter of `basicConsume()`:

```
ActualConsumer consumer = new ActualConsumer(channel);
boolean autoAck = false; // n.b.
channel.basicConsume(MyQueue, autoAck, consumer);
```

3. Consume a message and send the `ack`:

```
public void handleDelivery(String consumerTag,Envelope
    envelope, BasicPropertiesproperties,byte[] body) throws
        java.io.IOException {

String message = new String(body);
this.getChannel().basicAck(envelope.getDeliveryTag(),false)
    ;
```

How it works...

After we created the queue (step 1), we added the consumer to the queue and defined the `ack` behavior (step 2).

The parameter `autoack = false` informs the RabbitMQ client API that we are going to send explicit `ack` ourselves.

After we have got a message from the queue, we must acknowledge to RabbitMQ that we have received and properly processed the message calling `channel.basicAck()` (step 3). The message will be removed from the queue only when RabbitMQ receives the `ack`.

 If you don't send the `ack` back, the consumer continues to fetch subsequent messages; however, when you disconnect the consumer, all the messages will still be in the queue. Messages are not consumed until RabbitMQ receives the corresponding `ack`. Try to comment out the `basicAck()` call in the example to experiment this behavior.

The method `channel.basicAck()` has two parameters:

▶ `deliveryTag`
▶ `multiple`

The `deliveryTag` parameter is a value assigned by the server to the message, which you can retrieve using `delivery.getEnvelope().getDeliveryTag()`.

If `multiple` is set to `false` the client acknowledges only the message of the `deliveryTag` parameter, otherwise the client acknowledges all the messages until this last one. This flag allows us to optimize consuming messages by sending `ack` to RabbitMQ on a block of messages instead of for each one.

 A message must be acknowledged only once; if you try to acknowledge the same message more than once, the method raises a `precondition-failed` exception.

Calling `channel.basicAck(0,true)` all the unacknowledged messages get acknowledged; the 0 stands for "all the messages".

Furthermore, calling `channel.basicAck(0,false)` raises an exception.

There's more...

In the next chapter we will discuss the `basicReject()` method. This method is a RabbitMQ extension that allows further flexibility.

See also

The *Distributing messages to many consumers* recipe is a real example that explains better explicit `ack` use.

Distributing messages to many consumers

In this example we are showing how to create a dynamic load balancer, and how to distribute messages to many consumers. We are going to create a file downloader.

You can find the source at `Chapter01/Recipe10/Java_10/`.

Getting ready

To use this recipe we need to set up the Java development environment as indicated in the *Introduction* section.

How to do it...

In order to let two or more RabbitMQ clients properly balance consuming messages, you need to follow the given steps:

1. Declare a named queue, and specify the `basicQos` as follows:

    ```
    channel.queueDeclare(myQueue, false, false, false,null);
    channel.basicQos(1);
    ```

2. Bind a consumer with explicit `ack`:

    ```
    channel.basicConsume(myQueue, false, consumer);
    ```
3. Send one or more messages using `channel.basicPublish()`.

4. Execute two or more consumers.

How it works...

The publisher sends a message with the URL to download:

```
String messageUrlToDownload=
  "http://www.rabbitmq.com/releases/rabbitmq-dotnet-
    client/v3.0.2/rabbitmq-dotnet-client-3.0.2-user-guide.pdf";
channel.basicPublish("",MyQueue,null,messageUrlToDownload.getBytes
  ());
```

The consumer gets the message and downloads the referenced URL:

```
System.out.println("Url to download:" + messageURL);
downloadUrl(messageURL);
```

Once the download is terminated, the consumer sends the `ack` back to the broker and is ready to download the next one:

```
getChannel().basicAck(envelope.getDeliveryTag(),false);
System.out.println("Ack sent!");
System.out.println("Wait for the next download...");
```

By default, messages are heavily prefetched. Messages are retrieved by the consumers in blocks, but are actually consumed and removed from the queue when the consumers send the `ack`, as already seen in the previous recipe.

On the other hand, using many consumers as in this recipe, the first one will prefetch the messages, and the other consumers started later won't find any available in the queue. In order to equally distribute the work among the active consumers, we need to use `channel.basicQos(1)`, specifying to prefetch just one message at a time.

You can find more information about load balancing in *Chapter 8, Performance Tuning for RabbitMQ*.

Using message properties

In this example we will show how an AMQP message is divided, and how to use **message properties**.

You can find the source at `Chapter01/Recipe11/Java_11/`.

Getting ready

To use this recipe you will need to set up the Java development environment as indicated in the *Introduction* section.

How to do it...

In order to access the message properties you need to perform the following steps:

1. Declare a queue:

    ```
    channel.queueDeclare(MyQueue, false, false, false,null);
    ```

2. Create a `BasicProperties` class:

    ```
    Map<String,Object>headerMap = new HashMap<String,
      Object>();
    headerMap.put("key1", "value1");
    headerMap.put("key2", new Integer(50) );
    headerMap.put("key3", new Boolean(false));
    headerMap.put("key4", "value4");

    BasicProperties messageProperties = new
      BasicProperties.Builder()
    .timestamp(new Date())
    .contentType("text/plain")
    .userId("guest")
    .appId("app id: 20")
    .deliveryMode(1)
    .priority(1)
    .headers(headerMap)
    .clusterId("cluster id: 1")
    .build();
    ```

3. Publish a message with basic properties:

```
channel.basicPublish("",myQueue,messageProperties,message.
  getBytes())
```

4. Consume a message and print the properties:

```
System.out.println("Property:" + properties.toString());
```

How it works...

The AMQP message (also called content) is divided into two parts:

- Content header
- Content body (as we have already seen in previous examples)

In step 2 we create a content header using `BasicProperties`:

```
Map<String,Object>headerMap = new HashMap<String, Object>();
BasicProperties messageProperties = new BasicProperties.Builder()
.timestamp(new Date())
.userId("guest")
.deliveryMode(1)
.priority(1)
.headers(headerMap)
.build();
```

With this object we have set up the following properties:

- `timestamp`: This is the message time stamp.
- `userId`: This is the broker with whom the user sends the message (by default, it is "guest"). In the next chapter we'll see the users' management.
- `deliveryMode`: If set to 1 the message is nonpersistent, if it is 2 the message is persistent (you can see the recipe *Connecting to the broker*).
- `priority`: This defines the message priority, which can be 0 to 9.
- `headers`: A `HashMap<String, Object>` header, you are free to use it to enter your custom fields.

 The RabbitMQ `BasicProperties` class is an AMQP content header implementation. The attribute of `BasicProperties` can be built using `BasicProperties.Builder()`

The header is ready and we can send a message using `channel.
basicPublish("",myQueue, messageProperties,message.getBytes())`, where `messageProperties` is the message header and `message` is the message body.

In step 4 the consumer gets a message:

```
public void handleDelivery(String consumerTag,Envelope envelope,
BasicProperties properties,byte[] body) throws java.io.IOException {
System.out.println("**********message header***************");
System.out.println("Message sent at:"+ properties.getTimestamp());
System.out.println("Message sent by user:"+ properties.getUserId());
System.out.println("Message sent by App:"+properties.getAppId());
System.out.println("all properties :" + properties.toString());
System.out.println("**********message body*************");
String message = new String(body);
System.out.println("Message Body:"+message);
}
```

The parameter `properties` contains the message header and `body` contains its body.

There's more...

Using message properties we can optimize the performance. Writing audit information or log information into the body is a typical error, because the consumer should parse the body to get them.

The body message must only contain application data (for example, a `Book` class), while the message properties can host other information related to the messaging mechanics or other implementation details.

For example, if the consumer wants to log when a message has been sent you can use the `timestamp` attribute, or if the consumer needs to distinguish a message according to a custom tag, you can put it in the `headers HashMap` property.

See also

The class `MessageProperties` contains some pre-built `BasicProperties` class for standard cases. Please check the official link at http://www.rabbitmq.com/releases//rabbitmq-java-client/current-javadoc/com/rabbitmq/client/MessageProperties.html

In this example we have just used some of the properties. You can get more information at http://www.rabbitmq.com/releases//rabbitmq-java-client/current-javadoc/com/rabbitmq/client/AMQP.BasicProperties.html

Messaging with transactions

In this example we will discuss how to use channel transactions. In the *Producing messages* recipe we have seen how to use a persistent message, but if the broker can't write the message to the disk, you can lose the message. With the AQMP transactions you can be sure that the message won't be lost.

You can find the source at `Chapter01/Recipe12/Java_12/`.

Getting ready

To use this recipe you will need to set up the Java development environment as indicated in the *Introduction* section.

How to do it...

You can use transactional messages by performing the following steps:

1. Create a persistent queue

    ```
    channel.queueDeclare(myQueue, true, false, false, null);
    ```

2. Set the channel to the transactional mode using:

    ```
    channel.txSelect();
    ```

3. Send the message to the queue and then commit the operation:

    ```
    channel.basicPublish("", myQueue,
      MessageProperties.PERSISTENT_TEXT_PLAIN,
        message.getBytes());
    channel.txCommit();
    ```

How it works...

After creating a persistent queue (step 1), we have set the channel in the transaction mode using the method `txSelect()` (step 2). Using `txCommit()` the message is stored in the queue and written to the disk; the message will then be delivered to the consumer(s).

The method `txSelect()` must be called at least once before `txCommit()` or `txRollback()`.

As in a DBMS you can use a rollback method. In the following case the message isn't stored or delivered:

```
channel.basicPublish("",myQueue,
  MessageProperties.PERSISTENT_TEXT_PLAIN ,message.getBytes());
channel.txRollback();
```

There's more...

The transactions can reduce the application's performance, because the broker doesn't cache the messages and the `tx` operations are synchronous.

See also

In the next chapter we will discuss the **publish confirm** plugin, which is a faster way to get the confirmation for the operations.

Handling unroutable messages

In this example we are showing how to manage unroutable messages. An unroutable message is a message without a destination. For example, a message sent to an exchange without any bound queue.

Unroutable messages are not similar to dead letter messages; the first are messages sent to an exchange without any suitable queue destination. The latter, on the other hand, reach a queue but are rejected because of an explicit consumer decision, expired TTL, or exceeded queue length limit. You can find the source at `Chapter01/Recipe13/Java_13/`.

Getting ready

To use this recipe you will need to set up the Java development environment as indicated in the *Introduction* section.

How to do it...

In order to handle unroutable messages, you need to perform the following steps:

1. First of all we need to implement the class `ReturnListener` and its interface:

   ```
   public class HandlingReturnListener implements ReturnListener
   @Override
     public void handleReturn...
   ```

2. Add the `HandlingReturnListener` class to the channel `ReturnListener`:

   ```
   channel.addReturnListener(new HandlingReturnListener());
   ```

3. Then create an exchange:

   ```
   channel.exchangeDeclare(myExchange, "direct", false, false,
     null);
   ```

4. And finally publish a mandatory message to the exchange:

```
boolean isMandatory = true;
channel.basicPublish(myExchange, "",isMandatory, null,
  message.getBytes());
```

How it works...

When we execute the publisher, the messages sent to `myExchange` won't reach any destination since it has no bound queues. However, these messages aren't, they are redirected to an internal queue. The `HandlingReturnListener` class will handle such messages using `handleReturn()`.

The `ReturnListener` class is bound to a publisher channel, and it will trap only its own unroutable messages.

You can also find a consumer in the source code example. Try also to execute the publisher and the consumer together, and then stop the consumer.

There's more...

If you don't set the channel `ReturnListener`, the unroutable messages are silently dropped by the broker. In case you want to be notified about the unroutable messages, it's important to set the mandatory flag to `true`; if `false`, the unroutable messages are dropped as well.

2
Going beyond the AMQP Standard

In this chapter we will cover:

- ▶ How to let messages expire
- ▶ How to let messages expire on specific queues
- ▶ How to let queues expire
- ▶ Managing rejected or expired messages
- ▶ Understanding the alternate exchange extension
- ▶ Understanding the validated user-ID extension
- ▶ Notifying consumers of queue failures
- ▶ Understanding the exchange-to-exchange extension
- ▶ Embedding message destinations within messages

Introduction

In this chapter, we are going to present some recipes on the RabbitMQ extensions.

These extensions are not in AMQP 0-9-1 standard and using them will break AMQP compatibility with other compliant brokers.

On the other hand , they have been included, sometime in a slightly different form, in AMQP 0-10 (`http://www.amqp.org/specification/0-10/amqp-org-download`), so there is an easy path to port them. Finally, they are often the efficient solution to optimization issues.

The examples in this chapter are going to be more realistic, for example, configuration parameters, such as `queue` and `exchange`, and routing key names will be defined in the `Constants` interface in the Java sources. In fact, a real application will follow this approach or will eventually read them from configuration files to share them among different applications.

However, we will not specify the `Constants` namespace in the following sections to improve shortness and readability.

How to let messages expire

In this recipe, we are going to see how to let messages expire. The recipe sources can be found at `Chapter02/Recipe01/Java/src/rmqexample` in the code bundle that contains the following three executable classes:

- ▶ `Producer.java`
- ▶ `Consumer.java`
- ▶ `GetOne.java`

Getting ready

In order to use this recipe, we need to set up the Java development environment, as indicated in the *Introduction* section of *Chapter 1, Working with AMQP*.

How to do it...

The core of this example is in the `Producer.java` file. To produce messages that expire with a given **Time-To-Live (TTL)**, we have to perform the following steps:

1. Create or declare an exchange to send messages and bind it to a queue, as seen in *Chapter 1, Working with AMQP*:

   ```
   channel.exchangeDeclare(exchange, "direct", false);
   channel.queueDeclare(queue, false, false, false, null);
   channel.queueBind(queue, exchange, routingKey);
   ```

2. Initialize the TTL as an optional message property as follows:

   ```
   BasicPropertiesmsgProperties = new
   BasicProperties.Builder().expiration("20000").build();
   ```

3. Publish the messages using the following code:

   ```
   channel.basicPublish(exchange, routingKey, msgProperties,
   statMsg.getBytes());
   ```

How it works...

In this example, the producer creates the exchange, a named queue, and binds them to each other; messages with expiration make sense when it is possible that the queue is not attached to any consumer.

Setting the expiration time, the TTL (set in milliseconds), just let RabbitMQ delete any message as soon as it expires, if not consumed by a client in time.

In our example, we suppose that an application publishes JVM resource statistics to the given queue. If there is a consumer bound, it will just get real-time data as usual.

Alternatively, if there is nothing attached, messages will expire in the queue after the given TTL. In this way, we are avoiding collecting too much data.

As soon as a consumer binds to the queue, it gets the previous (not expired) messages and the real-time ones.

To further experiment, you can run the `GetOne.java` file instead of the `Consumer.java` file. This lets you consume just one message at a time by calling `channel.basicGet()`.

 It's possible to use `channel.basicGet()` to inspect a queue without consuming messages. It's enough to invoke it by passing `false` to the second parameter, that is, the `autoAck` flag.

In the course of the experiments, we can monitor the RabbitMQ queue status issuing `rabbitmqctl list_queues`.

See also

The expired messages are lost by default but they can be rerouted to other destinations. Refer to the *Managing rejected or expired messages* recipe for more information.

How to let messages expire on specific queues

In this recipe, we show a second way to specify a message TTL. This time, it is not a property of the message itself but of the queue where the message is buffered. In this case, the producer simply publishes normal messages to the exchange, so it's possible to bind both a standard queue and a queue where messages expire.

To remark on this aspect, here it's the consumer that creates the customized queue. The producer is quite standard.

As in the preceding recipe, you can find the three sources at `Chapter02/Recipe02/Java/ src/rmqexample`.

Getting ready

To use this recipe, we need to set up the Java development environment, as indicated in the *Introduction* section of *Chapter 1, Working with AMQP*.

How to do it...

We are now showing the needed steps to create a queue with a specific message TTL. In our explanatory example, the following steps are performed in the `Consumer.java` file:

1. Create or declare the exchange as follows:

   ```
   channel.exchangeDeclare(exchange, "direct", false);
   ```

2. Create or declare the queue, specifying a 10,000 milliseconds timeout to the `x-message-ttl` optional argument as follows:

   ```
   Map<String, Object> arguments = new HashMap<String, Object>();
   arguments.put("x-message-ttl", 10000);
   channel.queueDeclare(queue, false, false, false, arguments);
   ```

3. Bind the queue to the exchange where the messages are going to come from:

   ```
   channel.queueBind(queue, exchange, routingKey);
   ```

How it works...

Again, in this example, we are supposed to have a producer that sends JVM statistics to RabbitMQ for eventual analysis. Eventual because the `Producer.java` file sends them to an exchange and messages will be lost if there are no consumers connected.

A consumer that wants to monitor and analyze those statistics has the following three choices:

- ▶ To bind with a temporary queue, invoking `channel.queueDeclare()` without arguments
- ▶ To bind with a non-autodelete, named queue
- ▶ To bind with a non-autodelete, named queue and specifying `x-message-ttl`, as shown in step 2.

In the first case, the consumer will get real-time statistics only but it won't be able to perform an analysis on any data sent when it's down.

In the second case, to let the consumer get messages sent when it's down, it can use a named queue (persistent too eventually). But if it is down for a long time, when restarted, it will have a huge backlog to recover from before it starts being effective as a monitor. So, it would probably just trash most of the old messages in the queue.

The third option, the argument of our recipe, is just this, but trashing old messages is performed by RabbitMQ itself with benefits for both the consumer and the broker itself.

There's more...

When setting a per-queue TTL, as seen in this recipe, messages are not dropped (or dead lettered, refer to the *Managing rejected or expired messages* recipe later in this chapter) as soon as the timeout arrives, but only when a consumer tries to consume them. At this point, these messages are silently dropped and the first ready, not expired message is sent to the consumer.

When using queue TTL, this is a minor difference, but using per-message TTL, as seen in the previous recipe, it's possible to have expired messages in the broker queue behind regular messages.

In this case, those expired messages still occupy resources (memory) and are counted in the broker statistics, until they won't reach the head of the queue.

See also

The expired messages can be recovered in this case too. Refer to the *Managing rejected or expired messages* recipe.

How to let queues expire

In this third case, the TTL is not related to messages anymore, but to queues. This case is a perfect fit to manage server restarts and updates. The RabbitMQ drops the queue as soon as the TTL has elapsed, that is, after the last consumer has stopped consuming messages.

As in the previous TTL-related recipes, you can find the `Producer.java`, `Consumer.java`, and `GetOne.java` files in `Chapter02/Recipe03/Java/src/rmqexample`.

Getting ready

To use this recipe, we need to set up the Java development environment as indicated in the *Introduction* section of *Chapter 1, Working with AMQP*.

How to do it...

As in the previous example, the extension regards only `Consumer.java`:

1. Create or declare the exchange using the following code:

   ```
   channel.exchangeDeclare(exchange, "direct", false);
   ```

2. Create or declare the queue, specifying a 30,000 milliseconds timeout to the `x-expires` optional argument as follows:

   ```
   Map<String, Object> arguments = new HashMap<String,
     Object>();
   arguments.put("x-expires", 30000);
   channel.queueDeclare(queue, false, false, false,
     arguments);
   ```

3. Bind the queue to the exchange where messages are going to come from as follows:

   ```
   channel.queueBind(queue, exchange, routingKey);
   ```

How it works...

When we run either the `Consumer.java` or `GetOne.java` file, the timed queue is created and until there is a consumer attached to the queue or we call `channel.basicGet()`, it will continue to exist.

Only when we stop both the operations for at least 30 seconds, the queue is dropped and all the messages that it contains are lost.

 The queue is dropped independently from the fact that there are producers publishing messages to it or not.

In the course of the experiments, we can monitor the RabbitMQ queue status issuing `rabbitmqctl list_queues` and see this happening.

So, we can imagine a scenario where we have a statistics analysis program that needs to be restarted for an update of the code itself. It will restart without losing any messages, since its named queues will have a longer timeout. Alternatively, if we stop it, its queues will be deleted after the given TTL and worthless messages won't be stored at all.

Managing rejected or expired messages

In this example, we show how to manage expired or rejected messages using dead letter exchanges. The dead letter exchange is a normal exchange where dead messages are redirected; if not specified, dead messages are just dropped by the broker.

You can find the source in `Chapter02/Recipe04/Java/src/rmqexample`, where you can find the following files:

- `Producer.java`
- `Consumer.java`

To try expired messages, you can use the first code alone that publishes messages with a given TTL, as shown in the *How to let messages expire on specific queues* recipe.

Once started, the consumer of the example will not allow the messages to expire but will reject all the messages, leading to dead messages as well.

Getting ready

To use this recipe, we need to set up the Java development environment, as indicated in the *Introduction* section of *Chapter 1, Working with AMQP*.

How to do it...

Complete the following steps to show how to manage expired or rejected messages using dead letter exchanges:

1. We create the work exchange and the dead letter exchange:

   ```
   channel.exchangeDeclare(Constants.exchange, "direct", false);
   channel.exchangeDeclare(Constants.exchange_dead_letter,
     "direct", false);
   ```

2. We create the queue using the `dead_letter` exchange and `x-message-ttl` arguments:

   ```
   arguments.put("x-message-ttl", 10000);
   arguments.put("x-dead-letter-
     exchange",exchange_dead_letter);
   channel.queueDeclare(queue, false, false, false,
     arguments);
   ```

3. Then, we bind `queue` as follows:

   ```
   channel.queueBind(queue, exchange, "");
   ```

4. We can finally send the messages to exchange using `channel.basicPublish()`.

5. To try the rejected messages, we have to configure a consumer, as we have seen in the previous examples, and reject the message using the following code:

```
basicReject(envelope.getDeliveryTag(), false);
```

How it works...

Let's start with the first scenario (using the producer alone): the expired messages. In step 1, we create two exchanges, the working exchange and the dead letter exchange. In step 2, we create the queue with the following two optional parameters:

- Message TTL using `arguments.put("x-message-ttl", 10000)`, as seen in the *How to let messages expire on specific queues* recipe.

- The dead letter exchange name using `arguments.put("x-dead-letter-exchange", exchange_dead_letter);`

As you can see, we simply add the configuration to the optional queue arguments. So, when the producer sends a message to `exchange`, it will be routed to the `queue` parameter. The message will expire after 10 seconds, after which it will be redirected to `exchange_dead_letter`.

 The dead letter is a normal exchange, so you can use any one for this purpose.

For the second scenario, the consumer of this recipe will reject messages. When this consumer gets a message, it sends back a **negative acknowledgement** (**nack**) using `basicReject()` and when the broker receives the nack, it redirects the message to `exchange_dead_letter`. By binding a queue to the dead letter exchange, you can manage these messages.

When a message is redirected to a dead letter queue, the broker modifies the header message and adds the following values in the `x-dead` key:

- `reason`: This denotes whether the queue is expired or rejected (with `requeue = false`)
- `queue`: This denotes the queue source, for example, `stat_queue_02/05`
- `time`: This denotes the date and time the message was dead lettered
- `exchange`: This denotes the exchange source, for example, `monitor_exchange_02/05`
- `routing-keys`: This denotes the original routing keys used to send the message

To see these values in action, you can use the `GetOneDeadLetterQ` class. This class creates a `queue_dead_letter` queue and binds it to `exchange_dead_letter`.

There's more...

You can also specify the dead letter routing key using `arguments.put("x-dead-letter-routing-key", "myroutingkey")`. This parameter replaces the original routing key. This means that you can redirect different messages with different routing keys to the same queue. That's great!

Understanding the alternate exchange extension

We have already seen how to deal with messages published to an exchange that do not reach any queue in the *Handling unroutable messages* recipe in *Chapter 1, Working with AMQP*. AMQP lets the producer be acknowledged on this condition that can eventually decide whether to deliver the messages again to a different destination.

With this extension, we can specify an alternate exchange in which the broker will route these messages, without any more intervention from the producer, as shown in the sources of this recipe in `Chapter02/Recipe05/Java/src/rmqexample`.

Getting ready

To use this recipe, we need to set up the Java development environment, as indicated in the *Introduction* section of *Chapter 1, Working with AMQP*.

How to do it...

In this recipe, we are declaring the alternate exchange in the `Producer.java` class.

1. Put the name of the exchange where to route messages without the destination, `alternateExchange`, in an optional argument map with the `"alternate-exchange"` key as follows:

    ```
    Map<String, Object> arguments = new HashMap<String,
        Object>();
    arguments.put("alternate-exchange", alternateExchange);
    ```

2. Declare the exchange to send messages by passing the `arguments` map as follows:

    ```
    channel.exchangeDeclare(exchange, "direct", false, false,
        arguments);
    ```

3. Declare the `alternateExchange` itself, already specified in step 1, as follows:

    ```
    channel.exchangeDeclare(alternateExchange, "direct",
        false);
    ```

4. Declare a standard persistent queue and bind it to the `alternateExchange` using the routing key `alertRK` routing key:

```
channel.queueDeclare(missingAlertQueue, true, false, false,
   null);
channel.queueBind(missingAlertQueue, alternateExchange,
   alertRK);
```

How it works...

In this example again, we use a producer that generates statistics, as in the previous examples. But this time, we have added the routing key to let the producer specify an importance level, namely `infoRK` or `alertRK` (randomly assigned in the example).

Consumers attach their temporary queues to the `exchange` parameter declared by the producer in step 2. If you run the producer and at least one consumer, no messages are lost and everything works regularly.

 Consumers must pass the same optional arguments in the declaration of exchanges and queues, or an exception will be raised.

But if there are no consumers listening, we don't want to lose alerts. That's why we have chosen to let the producer create `alternateExchange` (refer to step 3) and bind it to a persistent queue, `missingAlertQueue` (refer to step 4).

On running the producer alone, you will see alerts being stored here. The alternate exchange lets us route messages that would otherwise be lost. You can check it by invoking `rabbitmqctl list_queues` or by running `CheckAlerts.java`.

This last code lets us view the contents of the queue and the first message, but doesn't consume any one. Accomplishing this behavior is simple, that is, it's enough to avoid the fact that the RabbitMQ client sends the ack and messages are not consumed, just monitored.

Now, if we run the `Consumer.java` file again, it will get and consume the messages from `missingAlertQueue`. This is not automatic; we have chosen to let it get the messages from this queue.

We have accomplished this by creating a second instance of the actual consumer (`missingAlertConsumer`) and letting the same code consume messages from two different queues. If we wanted different behaviors in the handling of real-time messages and missing messages, we would have created a different actual consumer.

There's more...

In this example, steps 3 and 4 are optional. It's possible to specify the name of the alternate exchange when defining an exchange, but it is not mandatory that it exists or is bound to any queue.

If the alternate exchange doesn't exists, it's possible for the producer to be acknowledged on the lost messages by setting the mandatory flag, as seen in the *Handling unroutable messages* recipe in *Chapter 1, Working with AMQP*.

It is even possible that an alternate exchange has... an alternate exchange of its own! So, alternate exchanges can be chained and messages without a destination will try them in sequence, until a destination is found.

If at the end of the alternate exchange chain no destination is found, the message is lost and it is possible for the producer to be notified by setting the mandatory flag and specifying a proper `ReturnListener` parameter.

Understanding the validated user-ID extension

According to AMQP, when a consumer gets a message, it doesn't know the sender. Generally, consumers should not care about who produced the messages; that's good for the producer-consumer decoupling. However, sometimes it's necessary for authentication, and RabbitMQ provides the validated user-ID extension for this purpose.

In this example, we're simulating a book order using validated user-IDs. You can find the source code in `Chapter02/Recipe06/Java/src/rmqexample`.

Getting ready

To use this recipe, we need to set up the Java development environment, as indicated in the *Introduction* section of *Chapter 1, Working with AMQP*.

How to do it...

Complete the following steps to simulate a book order using validated user IDs:

1. Declare or use a persistent queue as follows:

    ```
    channel.queueDeclare(queue, true, false, false, null);
    ```

2. Send a message specifying the user ID in the message header, using a `BasicProperties` object:

```
BasicProperties messageProperties = new
    BasicProperties.Builder()
.timestamp(new Date())
.userId("guest");
channel.basicPublish("",queue, messageProperties,
    bookOrderMsg.getBytes());
```

3. The consumer gets the order and prints the order data and the message header as follows:

```
System.out.println("The message has been placed by
    "+properties.getUserId());
```

How it works...

When the user-ID is set, the RabbitMQ checks whether it's the same one used to open the connection. In this example, the user is `guest`, the RabbitMQ default user.

The consumer can access the sender user ID, invoking the `properties.getUserId()` method.

If you try to set the user ID specified in step 2 to something different from the current RabbitMQ user, `channel.basicPublish()` raises an exception.

 If you don't use the user-ID property, the user is not validated and the `properties.getUserId()` method returns `null`.

See also

To understand this example better, you should know the users and vhost management that we will treat in the next chapter.

In the next chapter, we will see how to improve the application security using SSL. By using just the user-ID property, we guarantee authentication but all the information is not encrypted and can be easily exploited.

Notifying the consumers of queue failures

Following the AMQP standard, consumers are not informed about the queue deletions. A consumer waiting for messages on a queue that is deleted will not receive any error condition and will wait there indefinitely.

However, the RabbitMQ client provides an extension that lets the consumer receive a `cancel` parameter in such a case: consumer cancel notifications. We are going to see it in the example, which you can find in `Chapter02/Recipe07/Java/src/rmqexample`.

Getting ready

To use this recipe, we need to set up the Java development environment as indicated in the *Introduction* section of *Chapter 1, Working with AMQP*.

How to do it...

To make this extension work, you just need to perform the following step:

1. Override the `handleCancel()` method of the customized consumer, which we derived from `com.rabbitmq.client.DefaultConsumer` (refer to `ActualConsumer.java`):

   ```
   public void handleCancel(String consumerTag) throws
     IOException {
     ...
   }
   ```

How it works...

In our example, we have chosen to implement a consumer (of log events) that must work only if the producer is present and the queue where it listens is created by the producer.

So, if the queue is not present, the `Consumer.java` file exits immediately with an error. This behavior is accomplished by calling `channel.queueDeclarePassive()`.

The `Producer.java` class creates the queue as it starts and deletes it when closed, calling `channel.queueDelete()`. If it's closed while the consumer is consuming messages, the consumer is immediately notified by the RabbitMQ client library that calls `handleCancel()` overridden in step 1 of our recipe.

Other than being explicitly cancelled by calling `channel.basicCancel()`, the consumer can be cancelled for any reason using `handleCancel()`. Only in this case, the RabbitMQ client library invokes a different method of the `Consumer` interface: `handleCancelOK()`.

There's more...

Consumer cancel notifications are an extension of the client library, and not the general of all the AMQP client libraries. A library that implements them must declare it as an optional capability (refer to http://www.rabbitmq.com/consumer-cancel. html#capabilities).

The RabbitMQ client library supports and declares such a capability to the broker.

See also

In case a node fails and it is part of a RabbitMQ cluster, the same thing happens: a client consuming from its queues is not informed, unless it has defined its own handleCancel() override. For more information on this, refer to *Chapter 6, Developing Scalable Applications*.

Understanding the exchange-to-exchange extension

By default, AMQP supports exchange-to-queue but it doesn't support exchange-to-exchange bindings. In this example, we show how to use the RabbitMQ exchange-to-exchange extension.

In this example, we will merge the messages coming from two different exchanges to a third one. You can find the source code in Chapter02/Recipe08/Java/src/rmqexample.

Getting ready

To use this recipe, we need to set up the Java development environment as indicated in the *Introduction* section of *Chapter 1, Working with AMQP*, and run the producers from the *Broadcasting messages* and *Working with message routing using topic exchanges* recipes.

How to do it...

Complete the following steps to use the RabbitMQ exchange-to-exchange extension:

1. Declare the exchange where we want to trace the messages from using the following code:

   ```
   channel.exchangeDeclare(exchange, "topic", false);
   ```

2. Let's bind the exchanges from other examples to the new exchange using exchangeBind():

   ```
   channel.exchangeBind(exchange,ref_exchange_c1_8,"#");
   channel.exchangeBind(exchange,ref_exchange_c1_6,"#");
   ```

3. Start the trace consumer:

```
TraceConsumer consumer = new TraceConsumer(channel);
String consumerTag = channel.basicConsume(myqueue, false,
    consumer);
```

How it works...

In step 1, we create a new exchange that we bound (step 2) to the exchanges:

- `ref_exchange_c1_6` (broadcasting messages) with `exchange`.
- `ref_exchange_c1_8` (working with message routing using `topic`) with `exchange`.

In step 3, the consumer can bind a queue to `exchange` and get all the messages indiscriminately.

The exchange-to-exchange extension works exactly the same as the exchange-to-queue binding and you can specify a routing key to filter the messages. In step 2, we get all messages using # (the match-all pattern routing key as seen in *Chapter 1, Working with AMQP*) as the routing key. By changing the routing key, you make a filter!

Embedding message destinations within messages

In this example, we show how to use a single publish to send a message with multiple routing keys. The standard AMQP doesn't provide this feature but luckily, RabbitMQ does so using a message properties header. This extension is called **sender-selected distribution**.

The behavior of the extension is similar to e-mail logics. It uses **Carbon Copy** (**CC**) and **Blind Carbon Copy** (**BCC**). This is the reason you will find CC and BCC consumers in the source code, located in `Chapter02/Recipe09/Java/src/rmqexample`:

- `Producer.java`
- `Consumer.java`
- `StatsConsumer.java`
- `CCStatsConsumer.java`
- `BCCStatsConsumer.java`

Getting ready

To use this recipe, we need to set up the Java development environment as indicated in the *Introduction* section of *Chapter 1, Working with AMQP*.

How to do it....

Complete the following steps to use a single publish to send a message with multiple routing keys:

1. Create or declare the exchange using the following code:

```
channel.exchangeDeclare(exchange, "direct", false);
```

2. Send a message specifying the standard, CC, and BCC routing keys using the optional header properties of the message:

```
List<String> ccList = new ArrayList<String>();
ccList.add(backup_alert_routing_key);
headerMap.put("CC", ccList);
List<String> ccList = new ArrayList<String>();
bccList.add(send_alert_routing_key);
headerMap.put("BCC", bccList);
BasicProperties messageProperties = new BasicProperties.Builder()
headers(headerMap)
build();
channel.basicPublish(exchange, alert_routing_key,
   messageProperties, statMsg.getBytes());
```

3. Bind three queues to the exchange using the following three routing keys:

```
channel.queueBind(myqueue,exchange, alert_routing_key);
channel.queueBind(myqueueCC_BK,exchange,
   backup_alert_routing_key);
channel.queueBind(myqueueBCC_SA,exchange,
   send_alert_routing_key);
```

4. Get messages using the three consumers.

How it works...

When the producer sends a message using the CC and BCC message properties, the broker copies the message in all the queues bound with a matching routing key.

In this example, the stat class is directly sent to exchange with the routing key alert_routing_key. It is also copied to myqueueCC_BK and myqueueBCC_SA using the routing information in the CC and BCC arguments.

As it happens with e-mails, the BCC information is removed from the message header by the broker before it is dispatched to the queues; you can see this behavior in the output of all our example consumers.

There's more...

Normally, AMQP does not change message headers . The BCC extension is an exception to this rule that allows a further

This extension can be useful to minimize the number of the messages sent to the broker. Without the extension, a producer can just send many different copies of the messages with different routing keys.

3
Managing RabbitMQ

In this chapter we will cover:

- ▶ Using vhosts
- ▶ Configuring users
- ▶ Using SSL
- ▶ Implementing client-side certificates
- ▶ Managing RabbitMQ from a browser
- ▶ Configuring RabbitMQ parameters
- ▶ Developing Python applications to monitor RabbitMQ
- ▶ Developing your own web applications to monitor RabbitMQ

Introduction

Once installed, RabbitMQ just works; it's really a zero configuration service. However, RabbitMQ has a lot of configuration options, which make it very flexible and able to work in different environments. In this chapter we will see how to change the configuration to meet the requirements of your application.

We will also begin to use the tools that we will treat in detail in the following chapters in order to show you how to monitor RabbitMQ with an application.

Using vhosts

With **virtual hosts** (**vhosts**), it is possible to have many different, independent virtual brokers within one single RabbitMQ instance. In this way, it is possible to use the same broker on the parts of many different applications without worrying about name clashes. This is the same approach used by web servers with virtual hosts.

You can find the simple Java example in the `Chapter03/Recipe01` directory, which is identical to the example in the first recipe of the book, except for the usage of the vhost.

Getting ready

To exercise this recipe, you just need to issue some commands at the Linux command prompt, that is, **RabbitMQ Command Prompt (sbindir)** in the Windows Start menu.

How to do it...

To create a new vhost, perform the following steps:

1. List available virtual hosts with the command:

   ```
   rabbitmqctl list_vhosts
   ```

2. Create a new virtual host `book_orders` by issuing the command:

   ```
   rabbitmqctl add_vhost book_orders
   ```

3. List its exchanges:

   ```
   rabbitmqctl list_exchanges -p book_orders
   ```

4. List user permissions:

   ```
   rabbitmqctl list_permissions
   rabbitmqctl list_permissions -p book_orders
   ```

5. Allow the user, `guest`, to access the `book_orders` vhost:

   ```
   rabbitmqctl set_permissions guest .* .* .* -p book_orders
   ```

6. Use it with the Java client library:

   ```
   factory.setVirtualHost("book_orders");
   ```

How it works...

Once installed, RabbitMQ has just the default vhost / defined, as we can easily check with the command in step 1.

We then create a new vhost by issuing the command `rabbitmqctl add_vhost` (step 2). After that, we must issue all the commands related to this new vhost by specifying it with the `-p` option. If this is omitted, the commands are applied to the default vhost.

The new vhost you have added cannot be used yet. A quick check via listing permissions (step 4) will show that the new vhost has no authorization to perform any action. Then we give the predefined user, `guest`, all the permissions in the context of the `book_orders` vhost, as we will see in the next recipe (step 5).

At this point, it is enough to specify the vhost to the connection factory (step 6) in order to let a RabbitMQ client connect to it instead of the default vhost.

Configuring users

RabbitMQ users are global to the broker instance, but each user can have his/her own set of permissions for each individual vhost.

Different applications can be totally decoupled using independent users and vhosts.

However, the same application can benefit from the usage of user permissions within the same vhost.

We are going to see how to manage users and their permissions and how to use them in the Java example in `Chapter03/Recipe02`.

Getting ready

In order to run this recipe, we need to issue some `rabbitmqctl` configuration commands and exercise the configurations using the usual Java setup.

How to do it...

Perform the following steps to see how to manage users and their permissions as well as how to use them:

1. Create some users with their passwords:

   ```
   rabbitmqctl add_user stat_sender password1
   rabbitmqctl add_user stat_receiver password2
   ```

2. Grant some permissions to them:

   ```
   rabbitmqctl set_permissions stat_sender "stat_exchange.*"
       "stat_.*" "^$"
   rabbitmqctl set_permissions stat_receiver "stat_.*"
       "stat_queue_.*" "(stat_exchange_.*)|(stat_queue_.*)"
   ```

3. Let `Producer.java` connect to RabbitMQ using the `stat_sender` user:

```
factory.setUsername("stat_sender");
factory.setPassword("password1");
```

4. And then perform the following operations:

```
channel.exchangeDeclare(...);
channel.basicPublish(...);
```

5. Let `Consumer.java` connect using the user, `stat_receiver`:

```
factory.setUsername("stat_receiver");
factory.setPassword("password2");
```

6. Then perform the following operations:

```
channel.exchangeDeclare(...);
channel.queueDeclare(...);
channel.queueBind(...);
channel.basicConsume(...);
```

How it works...

In this exercise we have created a couple of users (step 1). In order to use them, it is needed to assign permission to them with the command:

```
rabbitmqctl set_permissions <username> <conf> <write> <read>
```

Here, `<conf>`, `<write>`, and `<read>` are three regular expressions. The specified `<username>` will have the permission to configure, write, and read queues and the exchanges matching them.

In our example, `Producer.java` accesses RabbitMQ with the `stat_sender` user. As shown in step 4, it has called `queueDeclare()`; so, it needs to have configuration permission for the exchange named `stat_exchange_03/02`.

It then publishes messages to the same exchange, so the user needs to have write permissions to it. But then messages will be routed to the `stat_queue_03/02` queue; the user also needs write permissions to it or the messages won't be routed to this queue. By setting the write permission to the regular expression `"stat_.*"`, the user is authorized to publish messages to both the exchange and the queue.

`Producer.java` does not need any read permission. It is possible to deny any read permission by specifying the empty regular expression `"^$"`, as shown in the example, or just an empty string.

On the other side, the user of `Consumer.java` needs the configure permission on both the exchange and the queue as well as the read permission in our recipe: `"stat_.*"` is synonymous to `"(stat_exchange_.*)|(stat_queue_.*)"` in the given example.

`Consumer.java` also needs write permissions on the `stat_queue_03/02` queue to be able to bind the given queue to the exchange.

There's more...

With users, we can restrict the access and capabilities of different RabbitMQ clients, tuning the set of permissions available to different components of a distributed application and conforming to their roles.

We can set a different set of permissions to each user, which applies to queues and exchanges specifying a pattern following regular expression matching rules.

Permission roles refer to the authorization to access specific AMQP resources:

▸ `configure`: The authorized user can declare and delete queues and exchanges that match the given pattern

▸ `write`: The authorized user can bind to queues and exchanges and publish messages to them if they match the given pattern

▸ `read`: The authorized user can bind to matching queues and exchanges and consume, get, and purge messages out of them

For the full table of the authorizations needed by the various AMQP operations, you can use the table at `http://www.rabbitmq.com/access-control.html`.

User tags for the management plugin

User permissions refer just to AMQP operations. The RabbitMQ management plugin (more details later in this chapter) extends this permission model using user tags; it's possible to associate one or more arbitrary tag strings to any user.

To let a user access the management plugin, it must have one of the following tags associated to it:

▸ `management`: The user has access to his/her vhosts only and can view all the queues and the exchanges within them

▸ `monitoring`: The user has access to all the vhosts and can view all the queues and exchanges

▸ `administration`: The user can do any administrative task

For example, to assign the administrator tag to a user, use the command:

rabbitmqctl set_user_tags username administrator

The default user, `guest`, has the `administrator` tag set by default.

Using SSL

Whenever the RabbitMQ broker is exposed to the Internet, it is highly advisable to protect the connections by using the SSL library.

RabbitMQ does not implement SSL by itself, but it uses the certificate mechanism of the given language, Erlang for the server and Java, .NET, or whatever for the clients.

Here, we will see how to have a basic protection with SSL, that is, how to encrypt the connections from the RabbitMQ clients to the broker.

This is enough to avoid simple security attacks. With no SSL, usernames and passwords are sent just in clear through the network.

Getting ready

The current example has the following prerequisites:

- ▶ A Linux OS hosting the RabbitMQ broker
- ▶ openssl Linux package
- ▶ The latest Erlang distribution—at least R14B
- ▶ Java JDK on the client, either on Linux or Windows

We have chosen to limit this recipe to just Linux on the server side because on Windows, there are too many version combinations—some with limited or no functionality at all. It is a good idea to run your secured Internet facing RabbitMQ servers on Linux or a Unix OS.

For more information on Windows support, you can go to `http://www.rabbitmq.com/ssl.html`.

How to do it...

Perform the following steps to set up the Certificate Authority and configure the server:

1. On the server, create the **Certificate Authority (CA)** directory stub as per the script in the example path `Chapter03/Recipe03/certificates/01_setup_CA.sh`:

    ```
    mkdir testca
    cd testca
    ```

```
mkdir certs private
chmod 700 private
echo 01 > serial
touch index.txt
```

2. Customize the `openssl` configuration file, which you can already find in `Chapter03/Recipe03/certificates/testca/openssl.cnf`.

3. Create the self-signed CA certificates as done in `Chapter03/Recipe03/certificates/02_create_CA_certificates.sh`:

```
openssl req -x509 -config openssl.cnf -newkeyrsa:2048
    -days 365 -out cacert.pem -outform PEM
       -subj /CN=MyTestCA/ -nodes
openssl x509 -in cacert.pem -out cacert.cer -outform DER
```

4. Create the RabbitMQ server private key as in `Chapter03/Recipe03/certificates/03_create_server_certificates.sh`:

```
openssl genrsa -out key.pem 2048
```

5. Create the server certificate request:

```
openssl req -new -key key.pem -out req.pem -outform PEM
    -subj /CN=$(hostname)/O=server/ -nodes
```

6. In the CA directory, sign the request to obtain the signed certificate:

```
openssl ca -config openssl.cnf -in ../server/req.pem -out
    ../server/cert.pem -notext -batch -extensions
       server_ca_extensions
```

7. Copy from `Chapter03/Recipe03/certificates`, the CA certificate, the server certificate, and the server private key, which we have just created, to the absolute paths:

```
/usr/local/certificates/testca/cacert.pem
/usr/local/certificates/server/cert.pem
/usr/local/certificates/server/key.pem
```

8. Create the RabbitMQ configuration file, `rabbitmq.config`, in the appropriate directory (`/etc/rabbitmq`) by copying it from `Chapter03/Recipe03/rabbitmq.config`:

```
[
{rabbit, [
{ssl_listeners, [5671]},
{ssl_options, [
{cacertfile, "/usr/local/certificates/testca/cacert.pem"},
{certfile, "/usr/local/certificates/server/cert.pem"},
```

```
{keyfile,"/usr/local/certificates/server/key.pem"},
{verify,verify_peer},
{fail_if_no_peer_cert,false}]}
]}
].
```

9. Restart the RabbitMQ server:

```
rabbitmqctl stop
rabbitmq-server-detached.
```

When you restart a RabbitMQ node, the Erlang Node will also restart.

10. In the Java client, the connection to the server is now made, as shown in `Chapter03/Recipe03/src/rmqexample/Publish.java`:

```
ConnectionFactory factory = new ConnectionFactory();
factory.setHost(hostname);
factory.setPort(5671);
factory.useSslProtocol();
Connection connection = factory.newConnection();
```

How it works...

We started this recipe by creating a CA with which we will sign the certificates for the server.

We have performed this step on the server, but in the real world, the CA and, in particular, its private key (created in step 3) should be kept separate.

After the CA is ready, we have to prepare the server certificate as shown in steps 4-6.

We are almost done. We just need to copy the CA certificate, the server certificate, and the server public key to the final path (step 7). We have chosen to store them in `/usr/local/certificates`, but it is totally arbitrary since they are referenced in the RabbitMQ configuration file, `rabbitmq.config`.

This file does not exist by default. It must be placed in the standard configuration directory, usually in `/etc/rabbitmq`.

Apart from the security files, we have configured the RabbitMQ SSL port (5671), and a couple of options in `rabbitmq.config`:

▶ `Verify`: When this is set to `verify_peer`, it tells RabbitMQ that if the client presents a certificate, it will be checked and rejected if not valid (because the CA is not the same or because it is expired)

▶ `fail_if_no_peer_cert`: When this is set to `false`, it tells RabbitMQ to accept clients that do not present any certificate at all

After we have restarted RabbitMQ (you must use `rabbitmqctl` stop and restart the service), you can verify whether it has got the options by examining the logfile in `/var/log/rabbitmq`. You should be able to find a line as follows:

```
started SSL Listener on [::]:5671
```

Furthermore, by opening the management plugin from a browser, you will be able to get similar information (see the *Managing RabbitMQ from a browser* recipe), as shown in the following screenshot:

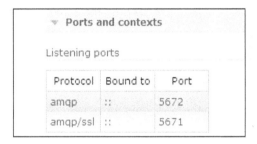

At this point it is possible to connect via SSL from a client by just adding these two options to the connection factory:

```
factory.setPort(5671);
factory.useSslProtocol();
```

Now the connection is encrypted using the keys configured in the server.

There's more...

Since we are using server certificates only, the communication between the server and the client is encrypted, but we are not protected against **MITM** (**man-in-the-middle**) attacks.

If we want to let any client connect to the server and avoid MITM attacks, we can use the same strategy as that used by HTTPS and the browsers, which is signing the certificates with trusted third-party CAs and verifying the domain signed in the certificates. Otherwise, we can just go on and read the next recipe.

Implementing client-side certificates

In case the RabbitMQ broker and client communicate through the Internet, it sounds reasonable that only authorized clients can connect to the broker.

This is the scope of typical user password authentication, but by using, in addition, client-side certificates, the security of the distributed application is highly improved. It also avoids the possibility of MITM attack.

This recipe is the extension/prosecution of the previous one. So, we assume that we already have the CA set up and the server configured, as shown previously.

Getting ready

This recipe is just an extension of the previous one—the same recommendations apply.

How to do it...

Perform the following steps for the client to be able to connect to the RabbitMQ server:

1. Copy the certificates and the keys, created in the previous recipe, from the `Chapter03/Recipe04/certificates` directory:

   ```
   cp -rp ../Recipe03/certificates/testca .
   cp -rp ../Recipe03/certificates/server .
   ```

2. In the client certificate directory, create the client private key, as shown in `Chapter03/Recipe04/certificates/04_create_client_certificates.sh`:

   ```
   openssl genrsa -out key.pem 2048
   ```

3. Create a certificate signing request:

   ```
   openssl req -new -key key.pem -out req.pem -outform PEM
   -subj /CN=$(hostname)/O=client/ -nodes
   ```

4. In the CA directory, sign the client certificate:

   ```
   openssl ca -config openssl.cnf -in ../client/req.pem
   -out ../client/cert.pem -notext -batch -extensions
     client_ca_extensions
   ```

5. Back in the client directory, create a `PKCS#12` store containing the client certificate and the key protected by a password:

   ```
   openssl pkcs12 -export -out keycert.p12 -in cert.pem
   -in keykey.pem -passout pass:client1234passwd
   ```

6. Create a Java key store containing the server certificate protected with a password, as shown in `Chapter03/Recipe04/certificates/05_create_keystore.sh`:

   ```
   keytool -importcert -alias server001 -file server/cert.pem
     -keystore keystore/rabbit.jks -keypass passwd1234
   ```

7. Change the `rabbitmq.config` option `fail_if_no_peer_cert` to true:

   `{fail_if_no_peer_cert,true}`

8. Restart RabbitMQ:

   ```
   rabbitmqctl stop
   rabbitmq-server -detached
   ```

9. On the client side, set up a secure connection by setting up the SSL context:

   ```
   char[] keyPassphrase = "client1234passwd".toCharArray();
   KeyStoreks = KeyStore.getInstance("PKCS12");
   ks.load(newFileInputStream("certificates/client/keycert.p12
     "), keyPassphrase);

   KeyManagerFactorykmf =
     KeyManagerFactory.getInstance("SunX509");
   kmf.init(ks, keyPassphrase);

   char[] trustPassphrase = "passwd1234".toCharArray();
   KeyStoretks = KeyStore.getInstance("JKS");
   tks.load(newFileInputStream("certificates/keystore/rabbit.
     jks"), trustPassphrase);

   TrustManagerFactorytmf =
     TrustManagerFactory.getInstance("SunX509");
   tmf.init(tks);

   SSLContext c = SSLContext.getInstance("SSLv3");
   c.init(kmf.getKeyManagers(), tmf.getTrustManagers(), null);

   ConnectionFactory factory = newConnectionFactory();
   factory.setHost(hostname);
   factory.setPort(5671);
   factory.useSslProtocol(c);
   Connection connection = factory.newConnection();
   ```

How it works...

For the client certificate to work, it must be signed with the same CA that has been used to sign the server. Once the certificate is prepared, it is very useful to save it in a keystore, a `PKCS#12` store as shown in step 5.

The client needs the server certificate too—it contains the server public key—and so we have prepared a keystore for this one too using a Java keystore with the Java keytool command this time.

Then we reconfigured RabbitMQ (steps 6-7). In this way, the server will deny access if the client has not presented any certificate. You can easily check this by running the previous example now.

Once the client certificates are ready, the RabbitMQ client (step 8) must use them to set up SSLContext. Unlike the previous example, we pass the following call:

```
factory.useSslProtocol(c);
```

Only a client following this setup can connect to the RabbitMQ server. The client presenting these certificates cannot connect to a server that has a different server private key since the traffic is being encrypted using the server public key stored in the server certificate.

Managing RabbitMQ from a browser

In this recipe we're showing you how to admin RabbitMQ from an HTTP API using a Management Plugin.

The plugin provides real-time charts to monitor the flow of your messages. Furthermore, it provides HTTP APIs to analyze RabbitMQ. This is required by external monitoring systems such as Ganglia (http://ganglia.sourceforge.net/), Puppet (http://puppetlabs.com), and others in order to perform their activities.

Getting ready

You just need a web browser.

How to do it...

In order to enable the plugin you have to perform the following steps:

1. Issue the following command:

    ```
    rabbitmq-plugins enable rabbitmq_management
    ```

2. Restart RabbitMQ.

3. The plugin enables a web server that is accessible via the URL http:// localhost:15672/. Replace localhost with your RabbitMQ hostname/IP to access from another machine.

How it works...

By default, the web application uses `guest/guest` as the RabbitMQ users' username/
password. Web management is very intuitive, and you can manage queues, exchanges, users,
connections, and virtual hosts and also send and receive messages.

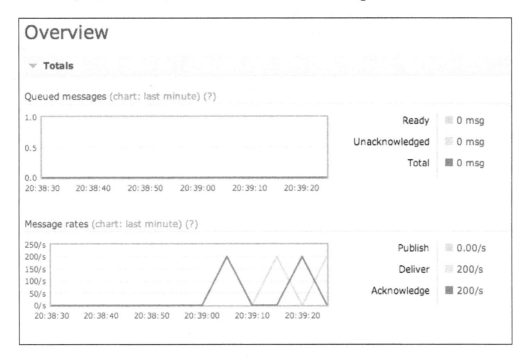

On the first tab, you can find the system overview with the queued messages, message rates,
and global count.

Then, you can find node description. Click on the node to get more details:

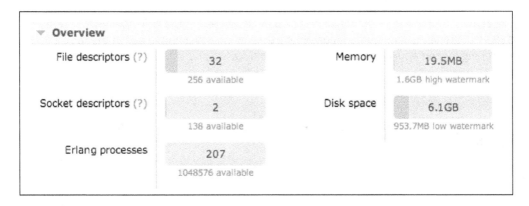

In the previous screenshot you can see some metrics that can help you to diagnose eventual problems.

> If you are running RabbitMQ on Windows, you need to install the `Hanlde.exe` SysInternals tool (`http://technet.microsoft.com/en-us/sysinternals/bb896655`) in an executable path such as `C:/Windows/system32`. Otherwise, the console won't be able to show the count of the file descriptors.

On the same page, there is other information about the active Erlang modules with their version numbers.

In the lower part of the overview, you can check the listening ports of the AMQP broker and the various installed plugins (web contexts):

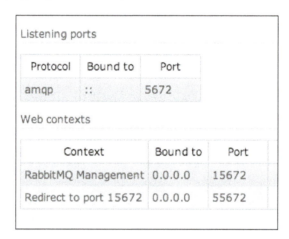

Listening ports

Protocol	Bound to	Port
amqp	::	5672

Web contexts

Context	Bound to	Port
RabbitMQ Management	0.0.0.0	15672
Redirect to port 15672	0.0.0.0	55672

There's more...

At `http://localhost:15672/api/`, you can get access to the HTTP API. By using the HTTP APIs, the users can create a custom console as we are going to see in the recipe *Developing Python applications to monitor RabbitMQ*.

See also

At `http://www.rabbitmq.com/management.html`, you can find the full documentation about the console.

Configuring RabbitMQ parameters

In this recipe, we are going to introduce the RabbitMQ parameters. By default, the broker doesn't create the configuration files because in most of the cases you don't need to change them. However, it's important to know how to configure the environment variables and the broker parameters.

How to do it...

In RabbitMQ, you can configure the **environment variables** and the server file configuration. With the environment variables, you can change parameters such as the server port or the node name. There are two ways to change these variables:

1. Define the variables in your shell environment.
2. Create a file `rabbitmq-env.conf` located in `/etc/rabbitmq`.

If, for example, you want to change the RabbitMQ node name, you have to do the following:

1. Stop the server.
2. Either issue `exportRABBITMQ_NODENAME=mylittlerabbit` on the shell or insert the string `NODENAME=mylittlerabbit` in the file, `/etc/rabbitmq/rabbitmq-env.conf`.
3. Restart the broker.

How it works...

Originally, the web management shows the default node name, `rabbit@hostname`, as follows:

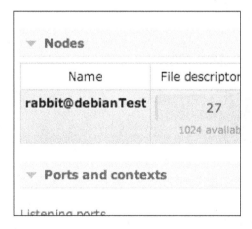

After our configuration, it shows `mylittlerabbit@hostname` as shown in the following screenshot:

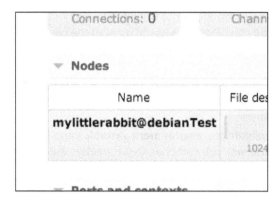

In this way, you can change all the parameters that you can find in the URL `https://www.rabbitmq.com/configure.html#define-environment-variables`.

> The environment variable name is prefixed with `RABBITMQ_`. You do not have to use it if you write the variable in `rabbitmq-env.conf`.

With the server file configuration, you can change the internal broker configuration. The configuration file is `rabbitmq.config` and the location is the same as that of `rabbitmq-env.conf`. Use the `RABBITMQ_CONFIG_FILE` environment variable to change the location. You can find the complete list of parameters at `https://www.rabbitmq.com/configure.html#configuration-file`.

There's more...

In this recipe we have just introduced the RabbitMQ configuration. In the next chapters, we will change some of the default parameters to tune the performance or configure the cluster.

Developing Python applications to monitor RabbitMQ

In this example we're going to create a Python script to monitor RabbitMQ using the JSON API that you can access from the URL `http://localhost:15672/api/`.

The scope of this example is to create a custom Python script, which performs some checks and sends an e-mail if it detects some errors. You can find the source code at `Chapter03/Recipe07`.

Getting ready

You need Python 2.7+ and the management plugin enabled (as we have seen in the *Managing RabbitMQ from a browser* recipe).

How to do it...

Perform the following steps to create a Python script to monitor RabbitMQ using the JSON API:

1. Import the following libraries:

```
import sys
import urllib2,base64
import json
import logging
```

2. Get the required RabbitMQ information from the JSON API:

```
urllib2.Request("http://"+rabbitmqhost+":15672/api/nodes")
urllib2.Request("http://"+rabbitmqhost+":15672/api/
  connections")
urllib2.Request("http://"+rabbitmqhost+":15672/api/
  aliveness-test/%2f")
```

3. Load and evaluate the JSON result:

```
json.load(urllib2.urlopen(request));
```

4. Send an e-mail if there is some error (or a condition for which you want to be alerted):

```
if error_message != "":
  sendAlarmStatus(error_message);
```

How it works...

With the /api/nodes JSON call, we check the `running state` and the `memory alarm` for each RabbitMQ node. With /api/connections, we are monitoring the connected clients. The /api/aliveness-test/ JSON call creates a test queue and tries to send and receive a message on the default virtual host %2f (this URL encoding is needed when referring to the character "/" in an URL, with "/" being the default AMQP virtual host). The result is {"status": "ok"} if there are no errors.

The script needs to authenticate the API using the code:

```
base64.encodestring('%s:%s' % ('guest', 'guest'))
  request.add_header("Authorization", "Basic %s" % base64string);
```

You must use a RabbitMQ user with monitor rights.

If some checks fail, the script sends an e-mail with the identified problems. To try this example, you need to modify the function `sendAlarmStatus(message)` with your e-mail account credentials.

 The script can be scheduled with the Linux crontab or using Scheduler Task in Windows.

Our example writes the logs in the `/var/tmp/myMonitorRMQ.log` file.

There's more...

In this example we have checked just a few APIs. Obviously there are more, which let you create a monitoring tool to check the RabbitMQ health customized depending on your requirements.

You can also use `rabbitmqadmin` to monitor the broker: the file is downloadable from the broker itself at `http://localhost:15672/cli/`. This is a Python script that can be invoked from a shell script, for example:

```
./rabbitmqadmin -f raw_json list nodes
```

This command will return the node information in the JSON format to the stdout, as we have already seen in step 2 of this recipe.

See also

You can find the whole API documentation at `http://hg.rabbitmq.com/rabbitmq-management/raw-file/rabbitmq_v3_0_4/priv/www/api/index.html`.

Developing your own web applications to monitor RabbitMQ

In this recipe we'll show you how to create a custom web application to monitor the RabbitMQ logs. In order to check the logs, you can bind a queue to the `amq.rabbitmq.log` exchange, as we will see in *Chapter 12, Managing RabbitMQ Error Conditions*, with much more details.

You can find the source code for this recipe in `Chapter03/Recipe08`.

Getting ready

In order to understand this example, we will make use of the following:

- Spring
- Apache Tomcat
- Apache Maven
- WebSocket
- Query/HTML5/Twitter Bootstrap

We recommend the Spring tool suite. You will need Internet access for the Maven, bootstrap, and jQuery includes.

How to do it...

Perform the following steps to create a custom web application to monitor the RabbitMQ logs:

1. Create a basic **MVC (Model View Controller)** Spring project using the available template from the tool as shown in the following screenshot:

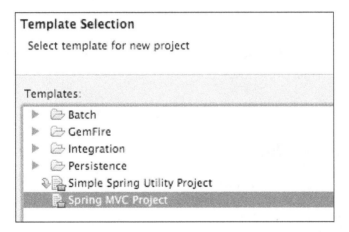

2. Modify the Maven file (POM.xml) by adding the following dependencies:

```
tomcat-coyote    // needed for websockets
tomcat-servlet-api// needed for websockets
tomcat-catalina// needed for websockets
spring-rabbit// needed for rabbitmq
```

3. Create a new bean `RabbitMQInstance` to create an interface with RabbitMQ using the `org.springframework.amqp` package.

   You can find the life cycle of bean in `root-context.xml`. In this example RabbitMQ must be run on the same host as Tomcat, but you can change the connection parameters.

4. Create a new class called `RmqWebSocket` that extends `WebSocketServlet` (Tomcat WebSocket class) and add the `@WebServlet ("/websocket")` annotation.

5. Create a private `amqp` consumer inside the `RmqWebSocket` class:

   ```
   LogListenerimplements MessageListener.
   ```

6. The consumer starts using `SimpleMessageListenerContainer` as follows:

   ```
   container.setConnectionFactory(rmq.getConnectionFactory());
   ...
   container.start();
   ```

7. Bind the browser WebSocket to the server using the code:

   ```
   ws = new WebSocket('ws://' host':8080/rmq/websocket');
   ```

8. Subscribe the browser using the code:

   ```
   ws.onmessage = function(message){..}
   ```

9. Each time the browser gets a message, it updates the HTML table using jQuery:

   ```
   $("#tablebody").prepend("<tr><td>"+message.data+
     "</td></tr>");
   ```

10. Now, we are ready. You can build Maven and deploy to the Tomcat container.

How it works...

In the bean `RabbitMQInstance`, we create and bind a queue to the topic exchange `amq.rabbitmq.log` with the routing key #. In the `root-context.xml` file, we define the bean ID and the init/deinit methods in order to connect AMQP clients during the web application startup and disconnect during the shutdown.

The RabbitMQ connection and queue are ready.

The `RmqWebSocket` is a `Tomcat websocket`, and here we start a consumer using the `SimpleMessageListenerContainer` implementation with the `rabbitMQInstance` parameters (connection and queue).

The consumer is ready.

With the method `StreamInbound` in `RmqWebSocket`, we register the `websocket` connection by the browser instantiating a new `ClientMessage` class for each connection and the `Collection<ClientMessage> clients` code line contains the active clients.

The `websocket` backend is ready.

In step 7 the browser is connected to the server and now it's ready, as shown in the following figure:

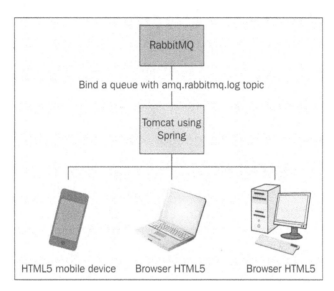

So when a log message is published to the topic exchange `amq.rabbitmq.log`, the class `LogListener` raises an event:

```
public void onMessage(Message message) {..}
```

The message is broadcasted to the connected browser with a `for` loop to the `clients` collection:

```
for(ClientMessage item: clients){
  CharBuffer buffer = CharBuffer.wrap(new
    String(message.getBody()) );
  item.myoutbound.writeTextMessage(buffer);
```

Now the message is received by the browser, which updates the local HTML table with `jquery`.

Tomcat, by default, uses the `8080` port and the URL application is `/rmq`. The complete URL is `http://yourtomcatmachine:8080/rmq/` and you should have a web browser pointing to this address as shown:

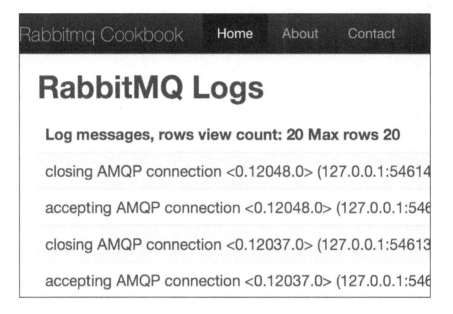

We have used HTML5 and Bootstrap, and you can also use a mobile device.

 The connecting or disconnecting of a client is enough to raise a standard log event.

There's more...

In this example, we have introduced many heterogeneous technologies in order to present a realistic example on how to merge RabbitMQ and current web technologies.

Here we have used RabbitMQ log messages, but you can easily apply the same architecture to any messaging need of your distributed/web application.

 In order to try to run the example directly, you can go to the source code path in the book archive (`Chapter03/Recipe08`) and prepare a package using Maven with **mvn package**. Then deploy the package to Tomcat.

See also

- Spring: http://spring.io/
- Apache Tomcat: http://tomcat.apache.org/
- Apache Maven: http://maven.apache.org/
- WebSocket: http://www.websocket.org/
- http://tomcat.apache.org/tomcat-7.0-doc/web-socket-howto.html
- jQuery: http://jquery.com/
- HTML5: http://en.wikipedia.org/wiki/HTML5
- Twitter Bootstrap: http://getbootstrap.com/

4
Mixing Different Technologies

In this chapter we will cover:

- ▸ Using a .NET client
- ▸ Binding an app from iPhone to RabbitMQ via MQTT
- ▸ Using messaging to update Google Maps on Android
- ▸ Publishing messages from Android in the background
- ▸ Exchanging RabbitMQ messages with Qpid
- ▸ Exchanging RabbitMQ messages with Mosquitto
- ▸ Binding a WCF application with .NET clients

Introduction

We have introduced the basic concepts in the previous chapters. Now, we will start using the concepts together in order to create real applications.

You won't find any new AMQP instruction, but you will see some typical usages of RabbitMQ; we will let RabbitMQ interact with other technologies to realize applications whose end users are both on desktop and mobile platforms. In fact the main objective of RabbitMQ/AMQP, and of messaging technologies in general, is to decouple different modules of distributed applications making it very easy to mix technologies with a wide spectrum of choices.

Using a .NET client

.NET is one of the official RabbitMQ clients, and in this example, we are going to create a subscriber using it. We will use **Microsoft Windows Presentation Foundation (WPF)** in order to create our first desktop application. Check out `http://msdn.microsoft.com/en-us/library/ms754130.aspx` for more information.

We are going to bind the .NET client to the Java dispatcher from the recipe *Broadcasting messages* in *Chapter 1, Working with AMQP*.

You can find the source code for this example in the directory `Chapter04/Recipe01`.

Getting ready

You need Visual Studio 2008 or later, Windows with .NET framework 3.5 or 4, and the RabbitMQ .NET client library, `RabbitMQ.Client.dll`, which you can download from `https://www.rabbitmq.com/dotnet.html`.

How to do it...

In order to realize a .NET application that makes use of RabbitMQ to consume messages, perform the following steps:

1. Create a WPF application and then add the `RabbitMQ.Client.dll` reference.

2. Connect to the broker using:

   ```
   connection_factory.Uri = "amqp://guest:guest@" +
       edRabbitmqHost.Text + ":5672/";
   ```

3. Bind a temporary queue to the `myLastnews.fanout_01/06` exchange:

   ```
   String myqueue = channel.QueueDeclare().QueueName;
   channel.QueueBind(myqueue,news_exchange, "");
   ```

4. Start a subscriber thread:

   ```
   sub = new Subscription(channel, myqueue, true);
   StartSubscriberThread(sub);
   ```

5. Receive a message using:

   ```
   foreach (BasicDeliverEventArgs e in sub) {
     ...
   Encoding.UTF8.GetString(e.Body);
   }
   ```

6. Write the message to the listbox:

```
Action<String> action = delegate(String value)
{lsnews.Items.Insert(0,value);};
Dispatcher.BeginInvoke(action,Encoding.UTF8.GetString(e.
    Body));
```

How it works...

The client is bound to the `myLastnews.fanout_01/06` exchange (created in the recipe *Broadcasting messages* in *Chapter 1, Working with AMQP*). Once a binding is created between the exchange (step 3) and the queue, we start a new thread using `StartSubscriberThread(sub)`, and the method executes `new Thread(() =>
InternalStartSubscriber(sub))` (step 4). The received message (step 5) is added to the listbox `lsnews` using `BeginInvoke()`.

In order to try the example, you have to do the following:

▸ Launch the publisher from `Chapter01/Recipe06/Java_6` using `java -cp ./
  rabbitmq-client.jar rmqexample.Publish`

▸ Launch the .NET client and connect to the broker

  ❑ The result should be as shown in the following screenshot:

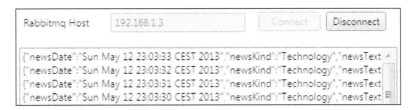

There's more...

One of the aims of AMQP is the integration among different languages, and now we have learned how to dispatch the same message with the following languages:

▸ Java

▸ Python

▸ Ruby

▸ .NET

  ❑ In the following recipes, we are going to add other clients and technologies.

See also

For the complete documentation, visit `https://www.rabbitmq.com/dotnet.html`.

Binding an app from iPhone to RabbitMQ via MQTT

MQTT (**Message Queue Telemetry Transport**) is an open source protocol, and is fast and lightweight. MQTT is widely used for the "Internet of Things". For more info, check out `http://mqtt.org/` and `http://en.wikipedia.org/wiki/Internet_of_Things`.

In this example, we are going to create an app that binds an iOS mobile device to RabbitMQ via MQTT using the Mosquitto library (`http://mosquitto.org/`).

We will see how to receive a message sent by:

- ▸ The RabbitMQ web management application presented in the *Managing RabbitMQ from a browser* recipe in *Chapter 3, Managing RabbitMQ*

- ▸ The publisher introduced in the *Working with message routing using topic exchanges* recipe in *Chapter 1, Working with AMQP*

You can find the source code for this example in the directory `Chapter04/Recipe02`.

 With MQTT, it's possible to send and receive messages, but it is not a replacement for Apple Push notifications.

Getting ready

To follow this recipe, you need:

- ▸ The Apple Xcode IDE (`https://developer.apple.com/xcode/`)

- ▸ The iOS developer library (`https://developer.apple.com/library/ios/navigation/`)

- ▸ The Mosquitto C library (`https://bitbucket.org/oojah/mosquitto/src`)

How to do it...

In this example, we need to enable the MQTT plugin to allow the MQTT clients' connections, and then create an iOS application by performing the following steps:

1. Install the MQTT plugin on RabbitMQ using `rabbitmq-plugins enable rabbitmq_mqtt`.

2. Restart RabbitMQ.

3. Create a new Xcode project without the **ARC (Automatic Reference Counting)** control as follows:

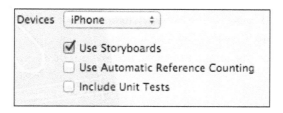

4. Download the Mosquitto C library (`https://bitbucket.org/oojah/mosquitto/src`) and from its `lib` directory, copy the `.h` and `.c` files. to the `libmosquitto` empty directory in the example tree.

5. Import the Moquitto C library and add the library to the project property `Header search path`.

6. Add the following `include` directive in your MQTT interface file:

    `#include "mosquitto.h"`

7. To start using the library, you need to initialize it by using `mosquitto_lib_init();`.

8. Create a struct `mosquitto *mosq;`, which you will use for each operation.

9. Initialize the structure using `mosq = mosquitto_new(NULL, true, self);`.

10. Connect to the RabbitMQ broker using the following code:

    ```
    mosquitto_username_pw_set(mosq, NULL, NULL);
    mosquitto_connect(mosq, cip, 1883, 60);
    ```

11. Subscribe to the queue using the following code:

    ```
    mosquitto_subscribe(mosq, NULL, "tecnology.rabbitmq.ebook", 0);
    ```

12. Set the message callback using `mosquitto_message_callback_set(mosq,on_message);`.

13. Start a loop to wait for messages from the server using `mosquitto_loop(mosq, 1, 1);`.

14. Your app is ready and you can send a message. First, you can use the web management console: go to the `amq.topic` exchange and publish a message using `technology.rabbitmq.ebook` as the routing key.

15. If you want to use the Java publisher from *Working with message routing using topic exchanges* recipe in *Chapter 1, Working with AMQP*, you have to bind the `myTopicExchangeRoutingKey_01/08` exchange with `amq.topic` using the routing key `technology.rabbitmq.ebook`, as we have seen in the *Understanding an exchange-to-exchange extension* recipe in *Chapter 2, Going beyond the AMQP Standard*.

16. Use `mosquitto_disconnect(mosq)` to disconnect the client, and `mosquitto_destroy(mosq)` to release the struct.

17. Finally, to release the library resources, use `mosquitto_lib_cleanup();`.

How it works...

Since version 3.0, RabbitMQ supports the MQTT protocol. You have to enable it using `rabbitmq-plugins`. Mosquitto is one of the most popular MQTT implementations, and in this example, we use its C client library. The plugin allows the communication between AMQP and MQTT!

After the initialization steps, and after following the steps 9-12, the client is ready; it means that the iPhone has created a queue bound to `amq.topic exchange` with `technology.rabbitmq.ebook` as the routing key. Now you can consider the queue as a normal AMQP queue.

Now open the web management console (the *Managing RabbitMQ from a browser* recipe in *Chapter 3, Managing RabbitMQ*) and send a test message (using the bound routing key) to see the example in action! The screen should look as follows:

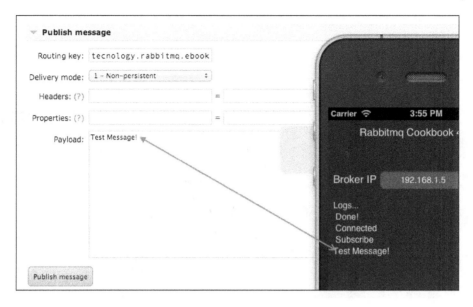

The test message has reached your iPhone!

Now suppose that the recipe *Working with message routing using topic exchanges* in *Chapter 1, Working with AMQP*, is part of a large enterprise system, which is difficult to be modified. You can't modify the publisher's and the consumer's software, but you need to redirect the "book's" info to the iPhone.

It's easy! It's enough to bind the exchanges from the management console using the e-2-e extension (the *Understanding an exchange-to-exchange extension* recipe in *Chapter 2, Going beyond the AMQP Standard*), setting the correct routing key in the **Add binding from this exchange** panel, as shown in the following screenshot:

The exchanges are bound. Now we can execute the publisher from the *Working with message routing using topic exchanges* recipe in *Chapter 1, Working with AMQP,* and see what happens! You'll see the messages correctly routed!

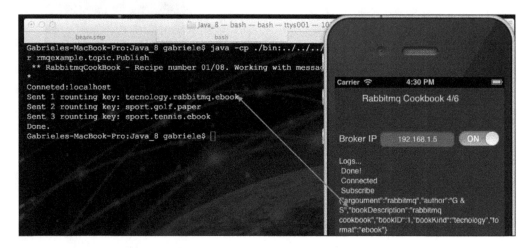

All the messages that are sent using `technology.rabbitmq.ebook` as the routing key will be redirected to the `amq.topic` exchange and then to your iPhone app.

There's more...

In this example, we have mixed the concepts studied in the previous chapters and some new technologies in order to touch on the various application contexts in which RabbitMQ can be used.

It's also important to understand how you can add or remove new modules in your existing application without modifying the software. It's great, isn't it?

 In case the Mosquitto server is outside the LAN, it's important to configure correctly the `keep-alive` parameter on the connection `mosquitto_connect(mosq, cip, 1883, 60 /*keepalive*/ )`.

See also

The MQTT website at `http://mqtt.org/`.

You need to read the full Mosquitto C client documentation at `http://mosquitto.org/api/files/mosquitto-h.html`.

You can run this example even on a MQTT server as `test.mosquitto.org` with the routing-key #.

Other interesting links are `http://www.eclipse.org/paho/` and `http://mqtt.org/2011/08/mqtt-used-by-facebook-messenger`.

Using messaging to update Google Maps on Android

Using RabbitMQ with Android is basically straightforward; the bytecode of the Dalvik virtual machine is compatible with the desktop Java versions, and the RabbitMQ JAR files containing the API are just the same.

In this recipe, we intend to plot a realistic example on how to use RabbitMQ in a mobile application, in particular, with the intent of drawing on Google Maps the position updates being published with the AMQP messages from another Android device.

However, this recipe is the first part of a two-recipe project; here, we focus mainly on the messaging aspects of the application, and in the next recipe, we will see how to put the messaging logic in an Android service.

The two recipes are also complementary, as in this recipe we are presenting a RabbitMQ consumer on Android, which plots the incoming positions on the map. The producer will be presented in the next one, and will publish its position through RabbitMQ. You can find the sources for this recipe in the directory `Chapter04/Recipe03`.

Getting ready

To put to work this recipe, we need the following:

- The Android SDK tools (`http://developer.android.com/sdk/index.html`)
- An Android device with Android API at least at level 11 (Honeycomb and Android 3.0.x or higher), and providing OpenGL ES 2.0 support
- Google Play installed on the Android device

How to do it...

1. From the Android SDK Manager, install Google Play Services:

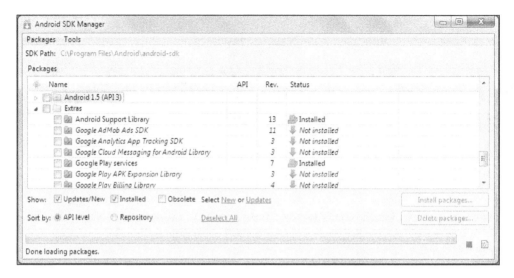

2. From within Eclipse, create a new Android application. Following the wizard, the minimum required SDK must be set to `API11`. Create a **Blank Activity**.

3. Carefully follow the instructions provided by Google itself to set up Google Play services, `http://developer.android.com/google/play-services/setup.html`, and to use the Google Map Android API v2, `https://developers.google.com/maps/documentation/android/start`.

4. Copy the RabbitMQ library files (`commons-cli-1.1.jar`, `commons-io-1.2.jar`, and `rabbitmq-client.jar`) to the `libs` directory of the project.

5. In place of the "Hello world!" TextView created by the Eclipse wizard, insert a fragment—you can edit the corresponding XML layout file `activity_main.xml` and insert the following:

```
<fragment
android:id="@+id/mapFragment"
android:name="com.google.android.gms.maps.MapFragment"
android:layout_width="match_parent"
android:layout_height="match_parent"
android:layout_alignParentLeft="true"
android:layout_alignParentTop="true" />
```

6. Create the simple class `MapCmd` containing the latitude and longitude of a position.

7. Create the singleton enum `MapController`, which implements a thread-safe FIFO queue of the `MapCmd` class.

8. Create the `ActualConsumer` class needed to consume RabbitMQ messages asynchronously as we have seen in *Chapter 1, Working with AMQP*, and override its `handleDelivery` method:

```
public void handleDelivery(
      String consumerTag,
      Envelope envelope,
      BasicProperties properties,
      byte[] body) throws java.io.IOException {
      String message = new String(body);
MapController.INSTANCE.AddCmdFromJson(message);
      }
```

9. Create the `RabbitmqHandler` class, exposing the `Connect()` and `Disconnect()` methods, which internally perform all the needed calls on the RabbitMQ API to let the app start consuming messages asynchronously. Customize the connection parameters included in this class as constants.

10. Create the `UpdateMap` class that is internal to the `MainActivity` instance of the `Runnable` interface, which periodically performs the drawing operations on the map:

```
public class UpdateMap implements Runnable {
  GoogleMapmMap;
  publicUpdateMap(GoogleMapmMap_) {
    mMap = mMap_;
    }
    @Override
public void run() {
    MapController.INSTANCE.ExecuteCmds(mMap);
    updateHandler.postDelayed(this, UPDATE_INTERVAL);
    }
  };
```

11. In the `MainActivity` class initially created by Eclipse, fill the `OnCreate()` method, using the building blocks created up to now:

```
protected void onCreate(Bundle savedInstanceState) {
    super.onCreate(savedInstanceState);
    setContentView(R.layout.activity_main);
    mMap = ((MapFragment)
      getFragmentManager().findFragmentById(R.id.mapFragment))
        .getMap();
    mMap.setMapType(GoogleMap.MAP_TYPE_HYBRID);

    updateMap = new UpdateMap(mMap);
    updateHandler = new Handler();
    updateHandler.postDelayed(updateMap, UPDATE_INTERVAL);
```

```
    rabbitmqHandler = new RabbitmqHandler();
    AsyncTask<Void, Void, Void>aconnect = new
      AsyncTask<Void, Void, Void>() {
      @Override
      protected Void doInBackground(Void... params) {
      try {
        rabbitmqHandler.Connect();
      } catch (IOException e) {
        // TODO handle this
      }
      return null;
    }
    };
    aconnect.execute((Void)null);
  }
```

12. Symmetrically implement the `OnDestroy()` method to clean up the connection and resources.

How it works...

In order to use Google Maps, Google Play should be installed so that its services are made available in both the Android SDK used for development, and on the device used to debug and deploy.

Furthermore, we must ensure that we have registered Google Maps in the Google API console (`https://code.google.com/apis/console`).

Don't forget to insert your own key in the `AndroidManifest.xml` file. In our example, we have left a placeholder:

```
<meta-data
android:name="com.google.android.maps.v2.API_KEY"
android:value="SET-YOUR-GOOGLE-ANDROID-MAPS-KEY-HERE"/>
```

After we have sketched a basic Google Maps application (steps 1-5), we have implemented the logic needed to receive the earth's coordinates from another peer, with the mediation of RabbitMQ.

Because of the need for responsiveness in an Android mobile application, it's mandatory that the receive operations are performed on a different thread from the main one, which must perform drawing and Maps-related operations.

That's why we have implemented a thread-safe FIFO queue of `MapCmd` objects (steps 6-7) and made it available in the `MapController` singleton; `MapCmd` objects are fed into this queue by the RabbitMQ consumer thread, calling `MapController.INSTANCE.AddCmdFromJson()`.

The consumer of the `MapController` FIFO queue is the main thread of the Android application. We have created a `Handler` object that periodically invokes the `updateMap.run()` method from the main thread, consumes `MapCmd` commands in the queue, and updates the displayed map accordingly.

To complete the applications, we must customize the `RabbitmqHandler` class with the IP address and the credentials of our RabbitMQ broker; this class handles the connection to RabbitMQ and the creation of the consumer threads.

We have created and initialized an instance of `RabbitmqHandler` in the main activity of the Android application (step 11), but it is worthwhile to note that we had to invoke its `Connect()` method from an `AsyncTask` anonymous object; it's mandatory to not perform any network-synchronous operation in the main thread—the connection to RabbitMQ included.

As we start the application on the Android device, we can see Google Maps loading the default worldwide map, but nothing more happens here; we must send a message.

To test this example, just open the RabbitMQ management console from a web browser (see the recipe *Managing RabbitMQ from a browser* in *Chapter 3, Managing RabbitMQ*); we should be able to see a temporary queue connected from the IP of the Android device. We can either publish messages directly to this queue or to the `amq.fanout` exchange, specifying the payload in JSON format, as shown in the following code example:

```
{"lat":44.4937,"lon":11.3430}
```

As we submit the message, the application will focus there. By sending more coordinates close to this one, we will get the route plotted as shown in the following screenshot:

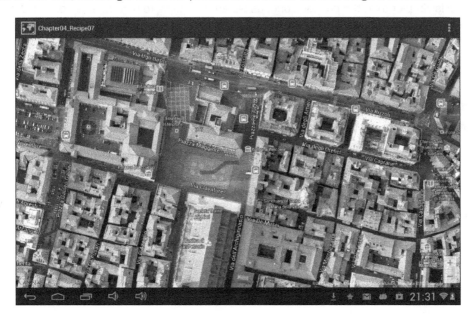

There's more...

In this recipe, we have shown how to develop a distributed application that communicates by using messaging on Android devices. For the sake of simplicity, we have omitted many details related to authentication, load balancing, security, and so on.

Furthermore, this example is just one half of a complete project; here, we have implemented the consumer only, publishing the JSON formatted messages from the RabbitMQ console.

The natural completion of the project shown here is in the next recipe, but before you jump there, it's worthwhile to note that the origin for the georeferenced data for this example is arbitrary—for example, it can be an embedded GPS device too.

 In case of embedded devices, and also for mobile, it would be better to use MQTT instead of the native RabbitMQ AMQP protocol for the sake of optimization, as we have seen in the *Binding an app from iPhone to RabbitMQ via MQTT* recipe.

Publishing messages from Android in the background

With this recipe, we are going to complete the project we began in the previous recipe; we will implement an Android app that publishes the current device's position to RabbitMQ. This example is simpler than the previous one, as we are not going to plot the positions on a map, but we still need the Google Play services since we have decided to make use of their location service; it provides a useful abstraction for the available location sensors available on the device, letting the application be compatible with a wide spectrum of Android devices.

Getting ready...

To put to work this recipe, we need:

▶ The Android SDK tools (http://developer.android.com/sdk/index.html)

▶ An Android device with Android API at least at level 11 (Honeycomb and Android 3.0.x or higher)

▶ Google Play installed on the Android device

How to do it...

1. From the Android SDK Manager, install Google Play services, as already seen in the previous recipe.

2. Create a new Android application. Following the wizard, the minimum required SDK must be set to `API11`. Create a Blank Activity.

3. Carefully follow the instructions provided by Google to set up the Google Play services (`http://developer.android.com/google/play-services/setup.html`).

4. Copy the RabbitMQ library files (`commons-cli-1.1.jar`, `commons-io-1.2.jar`, and `rabbitmq-client.jar`) in the `libs` directory of the project.

5. In place of the `Hello world!` TextView, created by the Eclipse wizard, insert a switch widget that we name `followmeSwitch`. You can edit the corresponding XML layout file `activity_main.xml` and insert the following:

```
<Switch
android:id="@+id/followmeSwitch"
android:layout_width="wrap_content"
android:layout_height="wrap_content"
android:layout_marginLeft="28dp"
android:text="@string/followme" />
```

6. Create the `SenderService` class that extends `IntentService`.

7. Set the action of the switch in `MainActivity.java`:

```
final Switch fms =
  (Switch)findViewById(R.id.followmeSwitch);
fms.setOnCheckedChangeListener(new
  CompoundButton.OnCheckedChangeListener() {
@Override
public void onCheckedChanged(CompoundButtonbuttonView,
  booleanisChecked) {
if (isChecked) {
serviceIntent = new Intent(MainActivity.this,
SenderService.class);
startService(serviceIntent);
} else {
stopService(serviceIntent);
}
}
});
```

How it works...

In this recipe, we have approached another typical pattern that we can meet by mixing RabbitMQ and Android technologies: publishing messages in the background, even when the user interface is not visible on the Android device.

We have first created a standard Android application with just a switch; on turning it on, we start delivering messages to the RabbitMQ broker. When turned off, we stop it.

So we have bound the actions of `followmeSwitch` to the `SenderService` (step 7). It's important to remember that when we start a background job with an Android service, the background events are still executed by the application's main thread, and we need to apply the same responsiveness guidelines that we follow when we design Android GUIs.

However, in our example, we have decided to make use of the `IntentService` helper class that starts a background thread that can run indefinitely.

It's in this thread that we connect to the broker and periodically send RabbitMQ messages reporting the current position of the device acquired by calling `locationClient.getLastLocation()` (see the source file with the class referred by the step 6).

> In this example, we have used the Android standard serializer `android.util.JsonWriter` in place of the RabbitMQ JSONWriter that we have seen previously. In fact, this one doesn't work on Android because Dalvik includes only a subset of the Java beans, missing the ones related to reflection.

As soon as the user turns off the switch, we just exit this thread.

By running this application on one device, and one of the previous recipes on another, we are able to communicate our georeferenced position by using RabbitMQ messages.

There's more...

The Google localization client library allows raising events on location changes, so this should be the preferred method to send coordinates. This approach can be optimized to conserve battery usage on the device, but we have avoided it for the sake of simplicity.

However, the application is already fully functional; you can run many different devices as consumers—and all of them will update the position synchronously, as we publish the messages to a fanout exchange.

You can even run many producers concurrently, but the current implementation will just mix their data together; to draw many different traces from different devices, it comes handy to replace the fanout exchange with a topic exchange and give a different routing key to each producer.

Exchanging RabbitMQ messages with Qpid

In this recipe, we will see how the RabbitMQ client can interoperate with Apache Qpid.

Getting ready

You need to download Qpid from `http://qpid.apache.org/download.html`; we have used the Java version.

How to do it...

1. Get a Java example from the previous chapter, for example, the *Broadcasting messages* or *Working with message routing using topic exchanges* recipe in *Chapter 1, Working with AMQP*.
2. Launch the Qpid server.
3. Execute the clients.

How it works...

When you execute the client from the example in the *Broadcasting messages* recipe in *Chapter 1, Working with AMQP*, the server has the same RabbitMQ behavior as you can see in the following screenshot:

QPid uses the same RabbitMQ port, so change the port or simply shut down RabbitMQ if you want to try both on the same machine.

There's more...

There are other AMQP brokers you can use for this test, but if you don't need to change the broker, you should always use the vendor client in order to take advantage of the broker's capabilities.

Exchanging RabbitMQ messages with Mosquitto

In this example, we are going to create a Java proxy that gets the messages published by the example presented in the recipe *Publishing messages from Android in the background* and forward them to any mobile OS as seen in the recipe *Binding an app from iPhone to RabbitMQ via MQTT*, using an MQTT server as follows:

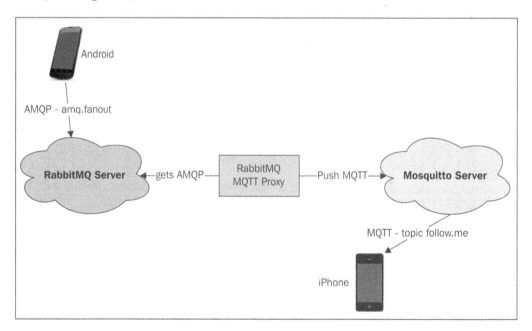

Getting ready

You need the Paho Java library client, which you can find at `http://git.eclipse.org/c/paho/org.eclipse.paho.mqtt.java.git/`.

How to do it...

1. You have to create a consumer for the `amq.fanout` exchange:

```
String myQueue = channel.queueDeclare().getQueue();
String myExchange = "amq.fanout";
channel.queueBind(myQueue,myExchange,"");
```

2. Create an MQTT connection with the Paho Java client:

```
mqttClient = new MqttClient(MQTTip,
  MqttClient.generateClientId());
mqttClient.connect();
```

3. Publish the retrieved GPS messages to the MQTT server:

```
mqttClient.publish("follow.me", message, 0, false);
```

4. Change the routing topic while binding the app from iPhone to RabbitMQ via MQTT:

```
mosquitto_subscribe(mosq, NULL, "follow.me", 0);
```

5. Change the server's IP while binding the app from an iPhone to RabbitMQ via MQTT, using the MQTT server IP.

How it works...

We have to create two connections, one to RabbitMQ and one to the MQTT server; follow the steps 1-3.

When the Android app sends some messages, the consumer recieves and publishes them to the Mosquitto server using `mqttClient.publish("follow.me", message, 0,false);`.

You need to change the topic on the iPhone app using `follow.me`, and you also have to change the server's IP address.

The result is as shown in the following screenshot:

So, if the Android app is enabled, you can see the latitude and longitude in your iPhone app.

We have used the public server `test.mosquitto.org`. Please read the site disclaimer before using it.

There's more...

We have divided the broker's responsibility, one for the AMQP protocol and one for the MQTT protocol.

See also

You can find more information about Paho at `http://www.eclipse.org/paho/`. When you build the Paho Java client source, the builder creates the Java doc in your filesystem: `yourbasedir/org.eclipse.paho.mqtt.java/org.eclipse.paho.client.mqttv3/build/dist/javadoc`.

Binding a WCF application with .Net clients

In this example, we will show how to create a distributed application that uses different technologies. The aim is to create a health application collector that receives the status from applications outside the local network. The application raises an alarm and sends it to the mobile phone if there is some problem; we have used Amazon EC2 as a cloud provider for the high availability. However, EC2 is not mandatory.

We are going to build an example using **Windows Communication Foundation** (**WCF**) and the RabbitMQ WCF service model. The WCF service will act as a "host health collector", and we will have one or more clients that send messages to the WCF service.

The RabbitMQ service model allows the SOAP protocol over AMQP. We also redirect the messages to a mobile iOS application via MQTT using the proxy already seen in the recipe *Exchanging RabbitMQ messages with Mosquitto*.

In this recipe, we are going to use AMQP, SOAP, and MQTT as protocols, RabbitMQ, WCF, and iOS as technologies, and finally, an AWS EC2 instance as the server. The architecture is summarized in the following figure:

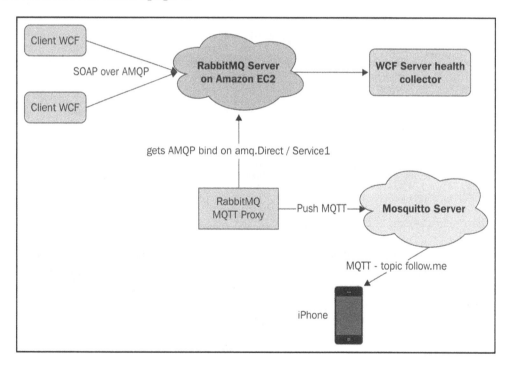

Getting ready

You need Visual Studio 2005 or 2008, .NET framework 3.5, and the RabbitMQ WCF service model, which you can find at `http://www.rabbitmq.com/dotnet.html`.

Download the zip archive with the WCF library.

How to do it...

1. Create a WCF application, and then configure the RabbitMQ service model by copying `RabbitMQ.ServiceModel.dll` in the `libs` directory on your project. Then open the `App.config` file with the right button and navigate to **Edit WCF configuration | Advanced | binding extension | new** and choose the file just copied.

2. Create a binding by navigating to **WCF configuration | bindings | new bind configuration**, and configure your RabbitMQ instance as shown in the following screenshot:

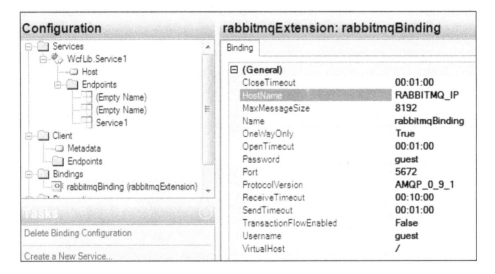

3. Configure the WCF end point by navigating to **Edit configuration | Endpoints | New Service End point**, and then set **Contract** to `WcfLib.IService1`, the **Binding** property to `rabbitmqExtension` (previously configured in step 1), and the **BindingConfiguration** property to `rabbitmqBinding` as (configured in step 2). The result will be as shown in the following screenshot:

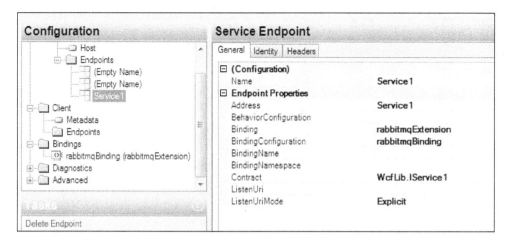

4. Configure the host by navigating to **Edit configuration | services | endpoint | new** and insert `soap.amqp:///`.

5. Create a `Service1` interface method `SendMachineInfo(MachineInfomachine Info);`, and implement it in the `Service1` class.

6. Create a new C# project and add the server dependency in order to use the interface methods.

7. Configure the client `App.conf` and add the configuration to the RabbitMQ service model (steps 1 and 2), and then configure the client service by navigating to **Edit configuration | Client | endpoints** and adding `soap.amqp:///Service1`; the binding will appear as shown in the following screenshot:

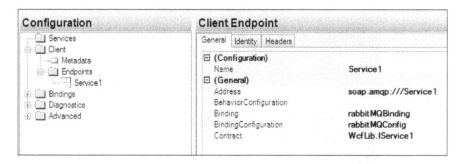

8. Start the client using `ClientService :ClientBase<IService1>, IService1`.

9. Call the interface using `base.Channel.SendMachineInfo(machineInfo);`

10. If you want to enjoy yourself, you can modify the recipe *Exchanging RabbitMQ messages with Mosquitto* in order to send the alarms to the iPhone. You just have to alter the following lines:

```
String myExchange = "amq.direct";
channel.queueBind(myQueue,myExchange,"/Service1");
```

How it works...

Following the steps 1-4, we extend the WCF standard binding with RabbitMQ, and we have created a **oneWay** communication. The oneWay is the "fire and forget" pattern, and the server creates only one queue bound to `amq.direct` with `/Service1` as the routing key.

The class `MachineInfo` is encoded in SOAP (step 9); the server can then execute the function `SendMachineInfo(MachineInfomachineInfo)`. The SOAP message is embedded as the payload of an AMQP message. In this example, the server stores the warning messages into the file `C:\warningstatus.store`, and the clients check the CPU and memory usage each second and execute the remote call in case of a warning. If you execute step 10, you should receive the messages to your app, shown as follows:

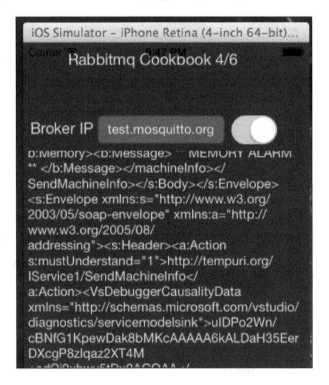

There's more...

We have tried this example using RabbitMQ on Amazon Elastic Compute Cloud (Amazon EC2) and both the WCF client and WCF server in different networks.

It's important to understand that the RPC calls are guaranteed by the broker, and are different from the standard RPC calls where the communication is normally direct between the modules. In this way, the architecture is decoupled, and RabbitMQ is used as a bus, where it's possible to attach different clients. The WCF binding is a powerful tool to integrate a Microsoft architecture with other systems.

You can monitor the queue using the web management console or the command line as a standard queue.

The communication can also be **twoWay**, using two queues: one for the requests and one for the asynchronous responses.

See also

Check out the documentation at `https://www.rabbitmq.com/dotnet.html` and find the section *A description of the WCF binding and Service Model*.

Check out `http://aws.amazon.com/ec2/` for more information about Amazon EC2; anyway, we will describe the Amazon cloud in the next chapters.

5
Using RabbitMQ in Web Applications

In this chapter we will cover:

- ▶ Developing web monitoring applications with Spring
- ▶ Developing asynchronous web searches with Spring
- ▶ Developing web monitoring applications with STOMP

Introduction

RabbitMQ can be used on the server side as well as on the client side. Currently, RabbitMQ covers the most used languages and technologies to build web applications, such as PHP, Node.js, Python, Ruby, and others. You can find the full list at `http://www.rabbitmq.com/devtools.html`.

In this chapter we will show three use cases, as follows, where RabbitMQ is used:

- ▶ It is used as a source for WebSockets' notifications (the *Developing web monitoring applications with Spring* recipe)
- ▶ It is used as a backend service to manage search requests (the *Developing asynchronous web searches with Spring* recipe)
- ▶ It is used directly on a web page to handle messages (the *Developing web monitoring applications with STOMP* recipe)

In this chapter we have chosen the Spring Framework to build most of the examples. We will show some interesting features, both for the integration and for RabbitMQ monitoring.

Developing web monitoring applications with Spring

In this recipe, we are going to show you how to build a web application to monitor the server CPU load and memory usage. The server status in the example is sent by a .NET client, but you can expand the example using other languages or operating systems; it's enough to publish a message using the de facto standard JSON protocol—a textual protocol with a very simple grammar (see `http://www.json.org/` for the full description). For example, a simple JSON-encoded message looks like the following text:

```
{"UPDATETIME":"23/06/2013
    22:55:32","SERVERID":"1","CPU":10,"MEM":40}
```

We will introduce Spring Insight now. With Insight, you can monitor the web application's performance and its correct behavior.

Check `http://gopivotal.com/products/pivotal-tc-server` for more information.

On the client side, we will use JQuery, Bootstrap by Twitter, and Google Chart.

At the end of the example, you will be able to put to work a working monitoring console where each server and the information is updated dynamically, as shown in the following screenshot:

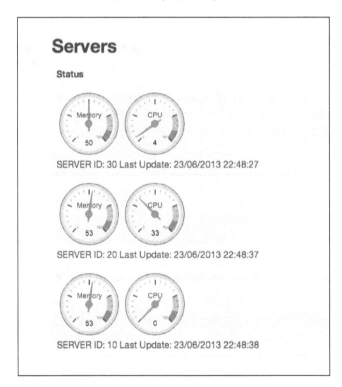

You can find the web application's source at `Chapter05/Recipe01/web` and the client at `Chapter05/Recipe01/csharp`.

Getting ready

To put the example to work, you need Java 1.7+.

In order to build the web application, we suggest you use Spring Tool Suite (also called STS) downloadable from `http://spring.io/tools`.

For the client, you need the .NET 3.5+ Framework.

How to do it...

To start quickly, you can use the MVC Spring template to create a complete and simple Spring project as follows:

1. Open STS and then navigate to **File | New | Other | Spring Template Project | Spring MVC Project**.

2. Modify `POM.xml` to add the Tomcat's `WebSocket` libraries:

   `tomcat-coyote`

   `tomcat-servlet-api`

   `tomcat-catalina`

3. Also, add the RabbitMQ client library to the POM:

   `amqp-client`

4. Create a `RabbitMQinstance` bean to store the RabbitMQ parameters.

5. Create the `RmqWebSocket` class that extends the Tomcat class, `WebSocketServlet`, in order to use the WebSocket. Inside the `RmqWebSocket.java` file, you will also find the RabbitMQ consumer code.

6. Create the `monitor_exchange_5_1` exchange and bind the consumer to it.

7. Implement the `handleDelivery` method from the `ActualConsumer` class to redirect the messages to the connected clients, iterating over the `clients` list as follows:

   ```
   for(ClientMessage item: clients){
   CharBuffer buffer = CharBuffer.wrap(message);
   try {
   item.myoutbound.writeTextMessage(buffer);
   ```

8. On the client side, where the JavaScript runs into the browser, connect the WebSocket instance using:

```
new WebSocket('ws://' + window.location.
    host  + window.location.pathname+ "websocket");
```

9. Implement the `ws.onmessage()` event to parse the JSON message and update the charts:

```
var obj = jQuery.parseJSON(message.data);
xcpu =obj.CPU ;
xmem = obj.MEM;
xupdate = obj.UPDATETIME;
var data = google.visualization.arrayToDataTable([
['Label', 'Value'],                        ['Memory',
    xmem],
['CPU', xcpu]]);
```

10. As a producer, we have created a .NET application in this example. In general, the producer has to publish a JSON message to the `monitor_exchange_5_1` exchange. The exchanged JSON message will look like the following code:

```
{"UPDATETIME":"23/06/2013
    22:55:32","SERVERID":"1","CPU":10,"MEM":40}
```

Here the application is ready, but if you want to configure Spring Insight, you have to perform the following steps:

11. Go to the **server** section, then click on **new server wizard**, and select **vFabric server** as shown in the following screenshot:

12. Then, move on and configure the instance name and the **insight** flag as follows:

13. Add your web application that you've created by following steps 1-10.

14. Start the vFabric server and go to the `http://localhost:8080/insight/` address.

15. In order to quickly install the RabbitMQ Insight plugin, you can download the JAR plugin from

    `http://maven.springframework.org/release/com/springsource/insight/plugins/insight-plugin-rabbitmq-client/`.

16. Copy the `insight-plugin-rabbitmq-client-XXX.jar` file in your `$STSHOMEINSTALLATION/vfabric-tc-server-developer-2.8.2/mytcinstance/insight/collection-plugins` folder.

17. Finally, copy the RabbitMQ Java client to `$STSHOMEINSTALLATION/vfabric-tc-server-developer-2.8.2/mytcinstance/insight/lib`.

18. Point your browser to `http://localhost:8080/example` to get the server load in real time, as shown in the screenshot reported at the beginning of the recipe.

19. Point your browser to `http://localhost:8080/insight` or click on the **Click if insight is enabled!** button on the home page to go to the Insight page.

How it works...

In steps 1-3 we have configured the environment; by default the wizard template creates a Maven project.

Spring 3.2.x doesn't support WebSocket yet (Version 4.x will), and we need to add the `websocket` Tomcat dependencies to the `POM.xml` file (step 2).

The RabbitMQ java client is also present on the Maven repository, and in step 3, we add the dependency.

The `RabbitMQinstance` bean contains the RabbitMQ connection parameters.

The bean's life cycle is defined in `root-context.xml`, with the `init` and `destroy` methods and custom parameters. (Check out `http://static.springsource.org/spring/docs/3.2.x/spring-framework-reference/html/beans.html#beans-factory-lifecycle` for more information.)

When the client sends a message (step 10) to the `monitor_exchange_5_1` exchange, the web application gets the message and redirects it to the WebSocket client's browser that is currently connected (step 7).

When the JSON message reaches the web page on the browser, it gets parsed by `JQuery` and finally it is used to update the Google Chart's pie diagrams.

> You have to deploy the application to Tomcat. If you use the STS standard server deploy, the application URL is as follows:
>
> `http://localhost:8080/example`

Steps 11-14 support the Spring Insight configuration, so you can monitor your application; using the RabbitMQ Insight plugin, you can also monitor RabbitMQ, as in the following screenshot:

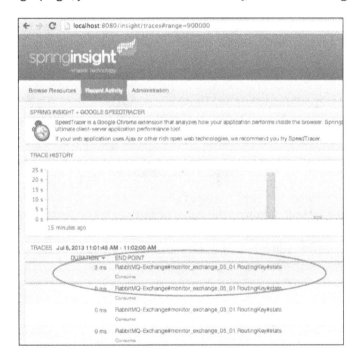

There's more...

Spring Insight can help you to monitor the application's behavior. Thanks to the plugin, you can also monitor RabbitMQ. It's certainly a more complex topic and we suggest you read from the following links:

```
http://gopivotal.com/products/pivotal-tc-server
```

```
http://blogs.vmware.com/management/2013/01/new-video-deep-dive-into-
spring-insights-plugins-for-spring-integration-and-rabbitmq.html
```

Anyway, we will still use it in the next recipes.

Developing asynchronous web searches with Spring

When developing a small site, usually the application server directly executes the query to a simple relational database and as a pattern, it is certainly very fast and easy to build. When the site grows, this pattern will show its scalability problems, especially when the server has to perform many operations to the query data before reporting back the result set that is to be presented to the end users.

Another problem is the system upgrade; each little update requires the application server to be stopped and restarted.

In this example, we are going to show how to use a message queue system to scale a web application. We will create separate modules with different responsibilities linked to each other with RabbitMQ.

The pros with respect to a traditional web application pattern (where there are only two modules: a web server and a database server connected to each other directly) are as follows:

- ▶ The communication is asynchronous
- ▶ Each module doesn't know the number and the network location of the other modules
- ▶ The additions, removals, and updates of the modules can be done seamlessly with respect to website responsiveness
- ▶ New modules with different scope can be added easily
- ▶ The architecture is scalable

The cons of the same are as follows:

- The application is more complex
- Locally, when compared to the simple and immediate approaches, the application could be slower

In this example, we are implementing an asynchronous, scalable, and balanced search engine to find an item (a book) with a keyword.

The following diagram explains the architecture that we are going to build:

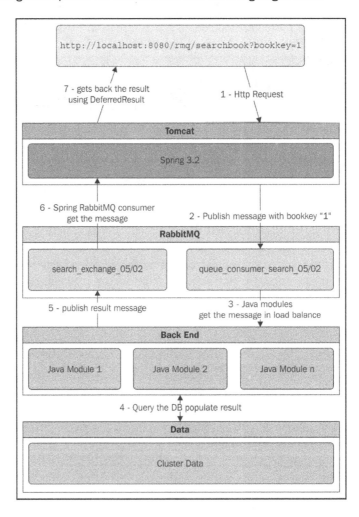

As you can see, we have divided the application's responsibilities as follows:

- ▸ Tomcat is just a proxy that takes the HTTP requests and publishes messages to the RabbitMQ broker

- ▸ The Java modules get the message in load balance, execute the needed queries to the DB, and return the results

- ▸ Tomcat gets the results back to the browser

In order to simplify the example, we have created a simple book list for each Java module to emulate a distributed database, where each module has a copy of the whole data. In the real case, you could have a clustered database or a clustered cache in front of a distributed database.

Different to the previous example, we use Spring instead of WebSockets to return the results: `DeferredResult` (`http://static.springsource.org/spring/docs/3.2.0.BUILD-SNAPSHOT/api/org/springframework/web/context/request/async/DeferredResult.html`).

You can find the web application source at `Chapter05/Recipe02/web` and the client at `Chapter05/Recipe01/backend`.

Getting ready

To put the example to work, you need Java 1.7+ and Apache Maven.

In order to build the web application, we suggest you use Spring Tool Suite (STS), downloadable from `http://spring.io/tools`.

How to do it...

We will skip the steps needed to create an MVC template application and the RabbitMQ configuration that we have presented in the previous recipe, and jump directly to the relevant topics of this one.

The following steps will show us how to implement the web module:

1. In the `RabbitMQInstance` bean, in the `Init()` method put the environment initialization, creating the `search_exchange_05/02` exchange and the `queue_consumer_search_05/02` queue.

2. Configure the message timeout:

```
Map<String, Object>args = new HashMap<String, Object>();
args.put("x-message-ttl", 4000);
channel.queueDeclare(Constants.queue, false, false,
    false, args);
```

3. Create an application UUID (http://en.wikipedia.org/wiki/Universally_
    unique_identifier) and use it as a routing key to subscribe the RabbitMQ client
    to the search_exchange_05/02 exchange:

```
channel.queueBind(myQueue,Constants.exchange,
    UUIDAPPLICATION);
```

4. In the HomeController class add the following method:

```
publicDeferredResult<String>searchbook(@RequestParam String
    bookid) {..}
```

5. Enable the asynchronous servlet using <async-supported>true</async-
    supported> in the web.xml file.

6. Register the HTTP calls to ConcurrentHashMap<String,
    DeferredResult<String>>.

7. Publish a message with the bookkey parameter to the queue_consumer_
    search_05/02 queue.

8. When the message reaches the SearchResultConsumer, there is a new result.

9. Find guidRequest on the requests map, and set the result to DeferredResult
    using setResult(message).

Let's continue the recipe by implementing the Java backend module as follows:

10. Subscribe to the queue_consumer_search_05/02 queue using channel.
    basicQos(1).

11. Handle the message and execute the search:

```
Book res= myDB.get(idx);
String jsonResult="";
    ..
jsonResult= jsonWriter.write(res);
```

12. Publish the result to the search_exchange_05/02 exchange as follows:

```
Builder bob = new Builder();
Map<String, Object> header = new HashMap<String,Object> ();
header.put("guidrequest", guidRequest);
bob.headers(header);
channel.basicPublish(Constants.exchange, RoutingKey,
    bob.build(), jsonResponse.getBytes());
```

How it works...

After we have created the environment (step 1), the web application is ready. We use the `queue_consumer_search_05/02` queue to put the query requests and the `search_exchange_05/02` exchange to get the query results.

In step 2, we have created an application UUID and used it as a routing key to the `search_exchange_05/02` exchange since you can have more Tomcat instances and more applications can query the backend. The unique UUIDs allow for distinguishing each module.

In step 3, we have implemented the GET handler for the `/searchbook` URL using `DeferredResult`. The `DeferredResult` class uses asynchronous support for servlet 3.0. (See `https://blogs.oracle.com/enterprisetechtips/entry/asynchronous_support_in_servlet_3` for more information.)

To enable the support for asynchronous servlet 3.0, we have modified the `web.xml` file (step 4).

With the `DeferredResult` class, you can set the result from a different thread without putting the current thread on hold. In this way, the incoming results will raise a callback enforcing event driver behavior at this layer too.

Let's see what happens for each HTTP request.

When the browser performs a request, for example, `/searchbook?bookkey=5`, the `searchbook` handler takes the request and assigns a new UUID to it. Then, it creates `DeferredResult<String>` and puts the pair into the `requests` map (step 6) just after the request parameters are published to `queue_consumer_search_05/02`. The message contains the request UUID and the application UUID:

```
Builder bob = new Builder();
Map<String, Object> header = new HashMap<String,Object> ();
header.put("guidrequest", guidRequest);
bob.headers(header);
bob.correlationId(UUIDAPPLICATION);
```

Now that the Java backend gets the message (step 11), it executes the query to the database (our emulated one with `localDbEmulation` in the example) and publishes the result to `search_exchange_05/02` (step 12).

In order to redirect the message correctly, the Java backend module sends back the request UUID and uses the `correlationId` header field as a routing key. In this way, we are sure that the response message is routed to the module that made the request.

In step 9, we have created the load balancing requests using the `channel.basicQos(1)` method (as already seen in the *Distributing messages to many* consumers recipe in *Chapter 1, Working with AMQP*). If you add more Java backend modules, they will automatically distribute the load among them.

So when the message result reaches the web application, it contains the following:

- The request UUID
- The JSON-formatted result

The `SearchResultConsumer` handles the message and finds the request in the `requests` map, and it also sets the result to `DeferredResult` as follows:

```
DeferredResult<String> deferredResult =
    requests.get(guidRequest.toString());
..
deferredResult.setResult(jsonResultSet);
```

The result is finally reported back to the browser, and the request is removed from the map.

```
requests.remove(guidRequest.toString());
```

The web and backend applications can be compiled using Maven. In this way, you can deploy the WAR and JAR files that it generates to more machines easily.

In this scenario, it's important to set a timeout for the operations in order to avoid the never-ending pending requests, and the `DeferredResult` class helps us with a constructor overload that takes the timeout in milliseconds as the argument:

```
new DeferredResult<String>(3000)
```

Even if the request goes in timeout, it must be removed from the `requests` map by defining the appropriate callback as follows:

```
deferredResult.onTimeout(new Runnable() {
public void run() {
rmq.requests.remove(guidRequest);
}
```

To avoid the pending messages in the queues, we have configured the message expire timeout using `x-message-ttl`. If a message remains in the queue for more than 4 seconds, it will be deleted (see *Chapter 2, Going beyond the AMQP Standard*). To make the example easy, we decided to just remove the long requests or pending messages—effective but drastic. In a real application, the timeout should be handled by reporting an alert to the administrators and presenting an "impossible to proceed" warning to the user.

See *Chapter 2, Going beyond the AMQP Standard* for the queue timeout; for `DeferredResult`, notify the error on its timeout handler:

```
deferredResult.onTimeout(new Runnable() {
public void run() {
rmq.requests.remove(guidRequest);
}
```

 Monitoring the timeouts is very important and can be a signal for the system to scale up your website. This can also be a sign that there is a problem.

To complete the example, we have used Apache JMeter (`http://jmeter.apache.org/`) in order to stress the website and see how much load it can sustain before we scale-up the application.

In the `Chapter05/Recipe02/jm/` folder, you can find the JMeter file that we have used to test the recipe, as you can see in the following screenshot:

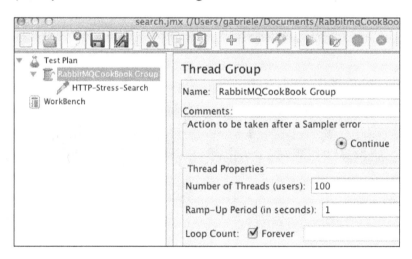

You can check that the application is behaving correctly by monitoring the log files or the STS console. If everything is running as expected, you will be able to see the following log line:

```
INFO :com.test.bean.RabbitMQInstance -  you have 0 pending requests
```

For the impatient, you can compile the examples with Maven using the mvn package on the root examples, then deploying the web WAR file, and executing the backend JAR file using `java -cp rmq-0.0.1-SNAPSHOT.jar:./rabbitmq-client.jar com.test.rmq.Main`. So, open the browser to `http://localhost:8080/rmq-1.0.0-BUILD-SNAPSHOT/` and make a search as shown in the following screenshot:

There's more...

In this recipe we have seen how a queue system is fundamental in an enterprise system. In this architecture, it's important to understand that you can scale easily by just adding more Tomcat instances or more Java backend modules, and as we will see in the next chapter, more nodes to the RabbitMQ broker depending on what is the bottleneck in the specific case.

The integration with a new kind of application is easy; it's enough to bind the new application to the exchange and to the queue.

It's also very easy to manage the updates of the software components with this architecture, avoiding any downtime. In fact, it's sufficient to have just one of the modules up in order to handle the requests.

See also

For more information, read from the following link:

`https://blogs.oracle.com/enterprisetechtips/entry/asynchronous_support_in_servlet_3`

Developing web monitoring applications with STOMP

In this recipe we will show you how to connect a web application directly to RabbitMQ using the Web-Stomp plugin. We will build the same application that we have seen in the *Developing web monitoring applications with Spring* recipe with the aim of having the same result with different technologies.

STOMP is Simple (or Streaming) Text Orientated Messaging Protocol (`http://stomp.github.io/`), which is another messaging wire protocol, alternative to AMQP and less powerful but with lighter clients.

The Web-stomp RabbitMQ plugin just provides the STOMP protocol interoperability over a SockJS server. On a different node, the STOMP plugin that we will see in *Chapter 9, Extending RabbitMQ Functionality* provides just plain STOMP protocol interoperability.

You can find the source at `Chapter05/Recipe03/HTML`.

Getting ready

You just need a text editor.

How to do it...

To cook this recipe, we are using the RabbitMQ plugin called Web-Stomp (`https://www.rabbitmq.com/web-stomp.html`), which is a STOMP bridge over emulated WebSockets, and we will access it using the JavaScript STOMP client.

1. Enable the plugin.

   ```
   rabbitmq-plugins enable rabbitmq_web_stomp
   ```

2. Restart RabbitMQ.

3. Test the plugin by opening the URL `http://127.0.0.1:15674/stomp` and you should get the **Welcome to SockJS!** message.

4. Create a simple HTML page and add the following library:

   ```
   <script src="http://cdn.sockjs.org/sockjs-0.3.min.js"></script>
   ```

5. Download `Stomp.js` from `https://raw.github.com/jmesnil/stomp-websocket/master/dist/stomp.js` and copy this file near the HTML file in `js/stomp.js`.

6. Add the `Stomp.js` reference to the HTML page:

   ```
   <script src="js/stomp.js"></script>
   ```

7. Initialize the connection parameter:

   ```
   var ws = new SockJS('http://' + window.location.hostname +
   ':15674/stomp');
   var client = Stomp.over(ws);
   ```

8. Connect the client to the RabbitMQ broker:

   ```
   client.connect('guest', 'guest', on_connect, on_error, '/');
   ```

9. Subscribe the client to the exchange:

   ```
   var on_connect = function() {
   client.subscribe("/exchange/monitor_exchange_05_01/
   stats",function(d) {
   ```

10. Parse the message and update the gauge:

```
var obj = jQuery.parseJSON(d.body);
xcpu = obj.CPU;
xmem = obj.MEM;
xupdate = obj.UPDATETIME;
```

How it works...

Once we have enabled and tested the plugin (steps 1-3), you can connect your JavaScript page to the RabbitMQ.

The first library you need (step 4) is the `Sockjs` JavaScript library (you can find more information at `http://sockjs.org`) and the second one is `Stomp.js`. `Stomp.js`, which is a JavaScript library for STOMP over WebSocket (also see `http://jmesnil.net/stomp-websocket/doc/`). After initializing the parameters (step 7), `client` can be connected to the RabbitMQ and then subscribed to the `on_connected` event.

The client is subscribed using the `subscribe` method (/exchange/monitor_exchange_05_01/stats), where the three slash-separated substrings are respectively stated as follows:

- ▸ The first substring could be of the type exchange, topic, or queue; in our example, `exchange`
- ▸ The second substring is the name; in our example, `monitor_exchange_05_01`
- ▸ The third substring is the routing key; in our example, `stats`

As in the first recipe, *Developing web monitoring applications with Spring*, you have to run the .NET client in order to send the status messages to the `monitor_exchange_05_01` exchange.

The outlook is identical to the one presented in the first recipe, and you can also execute the two recipes together (this and the first one). The messages will arrive at both the web applications, even if we are using substantially different technologies, between RabbitMQ and the browser.

 Different to the first recipe, to deploy the HTML page, you just need a web server like Apache.

The application developed in the course of this recipe is HTML5 compliant, and it's enough that the browser supports HTML5; you can use a mobile browser as well, as you can see in the following screenshot:

6
Developing Scalable Applications

In this chapter we will cover:

- ▸ Creating a localhost cluster
- ▸ Creating a simple cluster
- ▸ Adding a RabbitMQ cluster automatically
- ▸ Introducing a load balancer to consumers
- ▸ Creating clients of the cluster

Introduction

RabbitMQ provides, among the various features, clustering capabilities.

Using clustering, a group of properly configured hosts will behave the same as a single broker instance with the following purposes:

- ▸ **High availability**: If one node is shut down, the distributed broker still accepts and serves messages. These aspects will be treated in depth in *Chapter 7, Developing High-availability Applications*.
- ▸ **Scalability of the load of messages**: Message routing is distributed among all the nodes of the cluster.
- ▸ **Scalability of the connected clients**: Each node serves a subset of all the available clients.

Clustered RabbitMQ nodes should be in LAN. RabbitMQ clusters do not tolerate network partitions well because during a network partition each node works independently. There is no mechanism to prevent the so-called "split-brain syndrome" (find some good references to this problem at `http://en.wikipedia.org/wiki/Consensus_(computer_science)`. RabbitMQ uses short timeouts, typical of local networks, to determine the availability of sibling nodes. For more information, refer to `http://www.rabbitmq.com/nettick.html`.

All the nodes of a RabbitMQ cluster share the definition of vhosts, users, and exchanges but not queues. They physically reside on the node where they have been created. On the other side, they are globally defined and are reachable, connecting to any node of the cluster.

In case a client is producing or consuming messages from a node where the queue has not been created, performances will be lower because the client needs one more hop to reach the node physically hosting the queue.

 By default, queues are not replicated through the cluster, so in case of node failures, you can lose data. Refer to *Chapter 7, Developing High-availability Applications* to face this problem.

In this chapter, we will show some examples on the scalability aspects of RabbitMQ clustering and present the correct approach to design cluster-enabled AMQP clients.

Creating a localhost cluster

A localhost cluster is a RabbitMQ cluster where two or more instances of RabbitMQ are configured as a single cluster on the same host.

Actually, this is not very useful in production since it does not offer any scalability or reliability improvement. On the other hand, it is very useful to create a localhost cluster to test configurations on, for example, a development PC.

Getting ready

In order to put this recipe in action we just need RabbitMQ installed and running.

Even if it is not strictly required, we have installed the management plugin as shown in *Chapter 3, Managing RabbitMQ*. In this way, we need to specify a different web console TCP port for every broker.

How to do it...

In order to create a localhost cluster, we can perform the following steps:

1. From a root shell (Linux), start a second RabbitMQ instance with the following command:

   ```
   env RABBITMQ_NODENAME=node01 RABBITMQ_NODE_PORT=5673 RABBITMQ_
   SERVER_START_ARGS="-rabbitmq_management listener [{port,15673}]"
   rabbitmq-server -detached
   ```

2. Join this instance to the default one with the following commands:

   ```
   rabbitmqctl -n node01 stop_app
   rabbitmqctl -n node01 join_cluster rabbit@$HOSTNAME
   rabbitmqctl -n node01 start_app
   ```

3. Check your running cluster with the following command:

   ```
   rabbitmqctl cluster_status
   ```

4. Break the cluster back into individual nodes with the following commands:

   ```
   rabbitmqctl -n node01 stop_app
   rabbitmqctl -n node01 reset
   rabbitmqctl -n node01 start_app
   ```

How it works...

It is possible to run more than one instance of the RabbitMQ server on one machine, overriding some of the configuration options. In particular, it is mandatory to specify different ports for the broker itself (5672, by default) and for the management plugin port (15672, by default), if already installed.

This is accomplished by altering specific environment variables (refer to step 1). Currently, the available RabbitMQ environment variables are as follows:

- RABBITMQ_MNESIA_BASE
- RABBITMQ_LOG_BASE
- RABBITMQ_NODENAME
- RABBITMQ_NODE_IP_ADDRESS
- RABBITMQ_NODE_PORT

You can find the detailed description of all of the preceding variables in the `rabbitmq-server` manual page by typing the following command at the shell prompt:

man rabbitmq-server

Here, we are setting the RabbitMQ node name, its TCP port, and some custom arguments as follows:

RABBITMQ_NODENAME=node01

RABBITMQ_NODE_PORT=5673

RABBITMQ_SERVER_START_ARGS="-rabbitmq_management listener [{port,15673}]"

On Windows, the command issued at step 1 can be executed as a sequence of commands using the `set` command of the command line. You can copy the following small script in a `.bat` file for this purpose:

```
set RABBITMQ_NODENAME=%1
set RABBITMQ_NODE_PORT=%2
set RABBITMQ_SERVER_START_ARGS="-rabbitmq_management
listener [{port,%3}]"
"SBIN_FULL_PATH_HERE\rabbitmq-server" -detached
```

At this point, the second server that we have run is still independent of the first one. We must run the three commands shown in step 2, and then the two servers will be bound in a cluster.

On Windows, you have to use `%COMPUTERNAME%` in place of `$HOSTNAME`.

You can easily check this by running the examples in the book on this clustered environment. The exchanges created on the node and used by the code are replicated in the newly added node and vice versa. Just access the individual management plugins at `http://localhost:15672` and `http://localhost:15673` and see this.

Alternatively, you can issue the `rabbitmqctl` commands toward a specific node with the `-n` option, as performed in steps 2 and 4. If you do not use it, the command will be directed toward the node named `rabbitmq`, which is the default RabbitMQ server.

There's more...

The two nodes magically start speaking to each other, thanks to the fact that they share the same Erlang cookie since they are on the same machine.

In fact, different instances of RabbitMQ as well as of any Erlang-distributed application use a basic authentication scheme. Two instances of the application will be able to communicate to each other if and only if, they share the same Erlang cookie, which is just a file containing arbitrary data. It can be a string, a GUID, or whatever you think is the most appropriate.

The Erlang cookie is, in fact, a secret; if not properly protected, Erlang will not be allowed to use it and it won't work. You have to set its permissions properly, for example, with the following command:

```
chmod 400 .erlang.cookie
```

You can find more information on Erlang cookies at `http://www.erlang.org/doc/getting_started/conc_prog.html#id67454`.

It is possible that you will need to use the Erlang cookie, and it is mandatory to have the same one when we want to join cluster RabbitMQ servers running from different hosts, as we are going to see in the next recipe.

Creating a simple cluster

Setting up a cluster with RabbitMQ is a matter of a few minutes. With just a few configuration steps, the high-availability cluster is up and running.

It's actually easier than setting it up locally, as seen in the *Creating a localhost cluster* recipe, since each node will use the standard port settings.

Getting ready

In order to prepare this recipe, you need at least two hosts with RabbitMQ installed and configured as standalone brokers.

For the cluster to work properly, it's important that the versions of both RabbitMQ and Erlang are aligned on all the hosts.

The hosts can be physical servers, cloud instances, or virtual machines.

The **Amazon Web Services** (**AWS**) community AMI's with preinstalled RabbitMQ are usually outdated. We suggest installing a mainstream Linux AMI with your favorite distribution, and then installing the latest RabbitMQ version as per the installation guides (`http://www.rabbitmq.com/download.html`).

How to do it...

In the following steps, we will name the nodes as `node01`, `node02`, `node03`, and so on. Alternatively, you can use your current hostname, skipping steps 1 to 4.

1. Stop the RabbitMQ server on the nodes if it is already running:

   ```
   service rabbitmq-server stop
   ```

2. On `node01`, append all the IP-name bindings to `/etc/hosts`, so it contains something as follows (put your IP addresses here):

   ```
   10.0.0.1 node01
   10.0.0.2 node02
   ```
   ...

3. Copy the following file from `node01` to all the other nodes in order to have the hostname definitions aligned:

   ```
   scp /etc/hosts 10.0.0.2:/etc/hosts
   scp /etc/hosts 10.0.0.3:/etc/hosts
   ```
   ...

4. On all the nodes, set the local hostname (a different one on each server):

   ```
   echo node01 > /etc/hostname
   hostname -F /etc/hostname
   ```

   Then, we can configure RabbitMQ.

5. Copy the RabbitMQ Erlang cookie from `node01` to all the other nodes:

   ```
   scp /var/lib/rabbitmq/.erlang.cookie node02:/var/lib/rabbitmq/.erlang.cookie
   scp /var/lib/rabbitmq/.erlang.cookie node03:/var/lib/rabbitmq/.erlang.cookie
   ```
   ...

6. Restart the RabbitMQ server on all the nodes:

   ```
   service rabbitmq-server start
   ```

7. Join all the nodes to `node01`. On all the nodes except `node01`, run the following command:

   ```
   rabbitmqctl stop_app
   rabbitmqctl join_cluster --ram rabbit@node01
   rabbitmqctl start_app
   ```

At this point the cluster is up and running already. We can perform some more operations as follows:

1. Check the cluster status:

   ```
   rabbitmqctl cluster_status
   ```

2. Change the node type of node02:

   ```
   rabbitmqctl stop_app
   rabbitmqctl change_cluster_node_type disc
   rabbitmqctl stat_app
   ```

3. Let the cluster forget node02, which was invoked from node01:

   ```
   rabbitmqctl forget_cluster_node rabbit@node02
   ```

4. Let the removed node02 forget its clustered state. On node02 use the following commands:

   ```
   rabbitmqctl stop_app
   rabbitmqctl reset
   rabbitmqctl start_app
   ```

How it works...

It's important that all the servers in the cluster know each other by name since the IP address is not enough for RabbitMQ. So, you must configure and propagate /etc/hosts (or the equivalent hosts file in Windows) to all the servers.

 All the nodes must be addressable using short hostnames. If you try to use **fully qualified domain names** (**FQDN**), RabbitMQ won't work properly.

It's possible to use an external name server too, but in this case, the clustering would depend on it. It's usual practice to have the name server in front of the cluster and rely on static host information for the cluster configuration itself.

After all the nodes have been configured, all of them need to be synchronized, sharing the same cookie on all the servers. This file contains an arbitrary string and performs a very simple, mutual authentication among the nodes (step 5).

 Properly configuring the Erlang cookie lets rabbitmqctl work with remote nodes too, specifying them with the -n option.

At this point, we can create a cluster by issuing the following command:

```
rabbitmqctl join_cluster --ram rabbit@node01
```

If we have performed all the steps correctly, we will get the following output:

```
Clustering node rabbit@node02 with rabbit@node01 ...
...done.
```

By using the `--ram` option, we informed RabbitMQ that this node is a RAM node. No information is saved to the disk, making the node faster than a disk-based node (that is the default).

In the production of RabbitMQ clusters, at least a couple of nodes will be disc-based. In this way all the configuration (exchanges, queues, and so on) will be safe against reboots and failures of one of the disk nodes.

> Actually, since you have copied around the Erlang cookies, you can perform all these operations from `node01` eventually in a script, specifying the target node with the `-n` option:
>
> ```
> rabbitmqctl -n rabbit@node02 stop_app
> rabbitmqctl -n rabbit@node02 join_cluster rabbit@node01
> rabbitmqctl -n rabbit@node02 start_app
> ```

There's more...

In order to prepare this kind of configuration on a large cluster, it's common practice to use a distributed shell frontend.

Originally, `dsh` was only available in IBM AIX, but in modern Linux distributions you can find the useful `pdsh` and `pdcp` (usually in the `pdsh` package) commands performing the same operations.

You can use similar frontends on older distributions or on Windows too (such as the dancer's shell that can be found at `http://www.netfort.gr.jp/~dancer/software/dsh.html.en`).

Alternatively, you can perform the same operations with your favorite cluster management software.

See also

You can find detailed information in the RabbitMQ clustering guide available at `http://www.rabbitmq.com/clustering.html`.

Adding a RabbitMQ cluster automatically

Often, we need to prepare OS images that will join a RabbitMQ cluster once booted.

We can easily accomplish this task with some scripting, but RabbitMQ provides a simpler and more elegant option already, which we are going to explore with this recipe.

Getting ready

In order to prepare this recipe, you need at least two hosts with RabbitMQ installed and configured as standalone brokers.

The hosts must have the same version of RabbitMQ and Erlang.

How to do it...

In this recipe we start with two RabbitMQ core brokers that is a high-availability pair, which is the minimum condition for the cluster to work and be highly reliable. Then we configure all the other nodes as auto-configuring by writing the configuration to the `rabbitmq.config` file.

1. Configure the server hostnames as already shown in the *Creating a simple cluster* recipe steps 1 to 4.

2. Let all the servers have the same Erlang cookie.

3. Start RabbitMQ on `node01` and `node02` and create a cluster from them, as seen in the previous recipe.

4. If it has already been used, reset RabbitMQ on all the other nodes and halt it as follows:

   ```
   rabbitmqctl stop_app
   rabbitmqctl reset
   rabbitmqctl start_app
   ```

5. Set the content of the RabbitMQ configuration file, `rabbitmq.config`, on all the other nodes (the final dot is the part of the configuration file):

   ```
   [{rabbit,
     [{cluster_nodes, {['rabbit@node01', 'rabbit@node02'],
     ram}}]}].
   ```

6. Restart RabbitMQ on all the nodes as follows:

   ```
   service rabbitmq-server start
   ```

How it works...

This kind of configuration is the best suited to prepare and deploy OS images with RabbitMQ preinstalled. These images can be deployed and used to harmlessly grow an existing RabbitMQ cluster.

The `rabbitmq.config` file is read at the RabbitMQ startup. RabbitMQ will ignore the clustering autoconfiguration unless the node is in a totally clean state, which can be the case in either of the following two conditions:

 ▸ The RabbitMQ instance has been just reset and halted, as shown in step 4

 ▸ The RabbitMQ instance has been just installed and never started before

In the latter case, you don't need to perform step 6; just place the RabbitMQ configuration file in the right location, start RabbitMQ (which will happen in the automatic startup of the server if we are preparing an image), and the game is over.

> By default the RabbitMQ configuration file is named `rabbitmq.config` and will reside at different places, depending on the installation configuration. The following are the typical places:
>
> ▸ `/etc/rabbitmq/` (most distributions)
> ▸ `Rabbitmqdir/etc/rabbimq/` (Mac OSX)
> ▸ `%APPDATA%\RabbitMQ\` (Windows)

Once the node starts, it will repeatedly try to contact and connect to the nodes specified, as a list of nodes in the first element of the tuple of step 5. As soon as it succeeds with one of them, it will join to its cluster as a `ram` node, as specified in the second element of the tuple (or you can specify a `disc` node).

Introducing a load balancer to consumers

In order to have a RabbitMQ cluster with many nodes, the clients must know all the IP addresses, and if the cluster configuration is dynamic, the clients should be notified of any change.

Furthermore, the clients should employ some methods to contact the clients that are less loaded.

The common solution is to put a load balancer in front of the cluster.

A load balancer can be either a hardware solution or a software package, installed and configured on general purpose servers.

Furthermore, there are many different load balancing techniques, the following being the most relevant:

- ▸ Round-robin name servers
- ▸ TCP load balancers

TCP load balancers can behave as follows:

- ▸ **Proxy**: In this case, all the connections are just accepted and performed to the back and on part of the load balancer itself
- ▸ **Direct Server Return** (**DSR**): Accept just the connection initiation (SYN), and then let the connection be established directly between the client and the server in the backend without any further intervention (and with obvious benefits in the scalability and performance of the solution)

This is not all but going into further detail is out of the scope of this book. For our purpose, we will use a TCP software load balancer, Crossroads (`http://crossroads.e-tunity.com/`). After it is installed, the clients will connect with it, and it will forward the connection to the backend RabbitMQ servers, as shown in the following diagram:

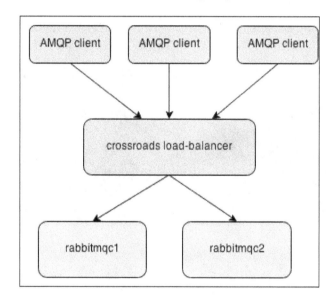

Getting ready

In this recipe, we need three machines; two will be clustered with RabbitMQ and one where we are going to install the load balancer.

In this case, the first two servers do not need to have public IP addresses if the clients are supposed to connect from the Internet. It's enough that the load balancer has one.

How to do it...

In this example, we have used two Linux Debian (with IP `192.168.10.2` and `192.168.10.3`) for RabbitMQ and one Linux Centos for Crossroads (with IP `192.168.10.1`).

Create a RabbitMQ cluster with two machines, as shown in the *Creating a simple cluster* recipe.

The following steps are only for the Centos machine (`192.168.10.1`):

1. In order to compile Crossroads, you need a C++ compiler and GNU make. You also need the system standard C libraries; execute the following command to prepare the machine:

   ```
   yum groupinstall "Development Tools"
   ```

2. Then, download the last Crossroads' stable version in your source directory (for example, `/usr/local/src/`), wget from `http://crossroads.e-tunity.com/downloads/crossroads-stable.tar.gz`.

3. Extract the `tar` file in your source directory using the following command:

   ```
   tar xvzf crossroads-stable.tar.gz
   ```

4. Navigate to the extracted directory `cd crossroads-2.74`, execute `make install`, and wait... at last you should have the following screenshot:

   ```
   The balancer program xr is now installed to /usr/sbin.
   The control script xrctl is installed there too. In order to
   use it, you will have to create /etc/xrctl.xml (if you have
   not done so yet). See "man xrctl.xml" for an example.

   Have fun with Crossroads 2.74,
       -- Karel Kubat <karel@kubat.nl>

   [root@rmloadbalance crossroads-2.74]#
   ```

5. Configure the load balancer for the RabbitMQ port:

   ```
   xr   -v --server tcp:0:5672 --backend 192.168.10.2:5672 --backend
   192.168.10.3:5672 2>&1 >> /var/log/xr-rmq.log &
   ```

6. If you have enabled the RabbitMQ web management plugin (Refer to *Chapter 3, Managing RabbitMQ*), you can also create a load balancer for it:

   ```
   xr -v --server tcp:0:80 --backend 192.168.10.2:15672 --backend
   192.168.10.3:15672 2>&1 >> /var/log/xr-web.log &
   ```

7. Open the browser on the URL `http://192.168.10.1`.

How it works...

After we have created the simple cluster (step 1), we have to install and configure the load balancer; for Crossroads, we suggest you to use a Centos Linux distribution. By following steps 2 to 5, you can compile and install Crossroads, and then you can configure the load-balancer roles. Firstly, we run the crossroads' load balancer with the following command (step 6):

```
xr  -v --server tcp:0:5672
--backend 192.168.10.2:5672
--backend 192.168.10.3:5672
2>&1 >> /var/log/xr-rmq.log &
```

The `xr -v server tcp:0:5672` command associates a server TCP listener with the `5672` port (the RabbitMQ default TCP port), the `--backend` parameter indicates the server and port to redirect the connection; in our case, the ports are `192.168.10.2:5672` and `192.168.10.3:5672`.

We use `>>/var/log/xr-rmq.log` to redirect the output to a logfile.

In order to access the web management plugin from the load balancer, we have to execute another XR server using the `15672` port, as in step 7.

Crossroads can use different dispatch algorithms but, by default, it uses the simple round-and robin algorithm, which means that the connections are equally redirected to each backend server.

At this point, you can run any RabbitMQ client example, replacing the server IP address with the crossroads IP. The connection will be automatically forwarded to the backend brokers. In the next chapter, we will see the load balancer in action using a client and other high availability policies.

There's more...

Crossroads is one of the easiest Linux load balancers; refer to `http://crossroads.e-tunity.com/documentation.xr` for the complete documentation. Another widespread load-balancing solution that you can use for this example is HAProxy (`http://haproxy.1wt.eu/`).

See also

Please read the section *Connecting to Clusters from Clients* from the URL `http://www.rabbitmq.com/clustering.html` for more details.

Creating clients of the cluster

When you create a simple cluster, the clients may need to specify more than one broker address when connecting. In this recipe, we are going to see how to use the multiple address connections with the RabbitMQ Java client.

Note that in case you have a load balancer in front of the cluster, as shown in the previous recipe, you don't need to use multiple address connections. The load balancer performs this work for you.

Getting ready

You need Java 1.6 or higher and Apache Maven.

How to do it...

Firstly, you need a simple cluster as seen in the *Creating a simple cluster* recipe.

Then you can connect your client using the following code:

```
Address[] addrArr = new Address[]{ new
   Address("node01",portnode1), new Address("node02", portnode2)};
connection = factory.newConnection(addrArr);
```

How it works...

With the RabbitMQ libraries, you can pass more than one host/IP to the connection factory. This feature can be used for small and big scenarios. For example, if you have two nodes and you want to avoid using a DNS or load balancer, you can pass all the broker addresses directly from the client. The client will try to connect to the first address. If the first address is down, the client will try with the next one without raising exceptions.

The same schema can be applied with a complex architecture. For example, suppose you have the following two hosts:

▸ myrmqcluster_production.internal.com

▸ myrmqcluster_maintenance.internal.com

The clients will always connect to the production system, but if the production system is in maintenance, the clients will connect to the second address.

There's more...

Actually, you can also use this connection method without a RabbitMQ cluster. In the source code that you can find at `Chapter06/Recipe06`, we create a dummy algorithm to balance the connections. In this case, you can have two independent RabbitMQ instances.

See also

In the next chapter, we will see in detail how to create an HA client by mixing together clusters, HA policies, and client technologies.

7
Developing High-availability Applications

In this chapter we will cover:

- ▶ Mirroring queues
- ▶ Synchronizing queues
- ▶ Optimizing mirror policies
- ▶ Distributing messages between a couple of brokers
- ▶ Creating a geographical cluster replication
- ▶ Filtering and forwarding messages
- ▶ Combining the high-availability technologies together
- ▶ Client high availability

Introduction

RabbitMQ approaches the issue of high availability by replicating data, the same as storage solutions (think of the RAID solutions), databases, and all the IT infrastructures when data integrity and service continuity are of primary importance.

In fact, the objective of these kinds of solutions is not only to avoid the possibility of data loss but also to avoid any downtime due to both scheduled maintenance and system malfunctions.

We will see how to use the simple but effective solution provided by RabbitMQ with queue mirroring. Through the recipes, we will see different use cases and optimizations that can be approached to minimize the performance price that you always have to pay when dealing with high availability.

Then, we will see how to perform geographical replication. This is an approach needed when the required **Quality of Service** (**QoS**) is so high that the application needs to be available even when a whole site is down (for example, due to a problem on the grid and the emergency power supply, a natural disaster, or just human error).

This approach is also suggested when using cloud computing resources. For example, when using **Amazon Web Services** (**AWS**), it is highly recommended to distribute the application among different availability zones (`http://docs.aws.amazon.com/AWSEC2/latest/UserGuide/using-regions-availability-zones.html`). Since the link among different zones has much higher latency, it would not be reasonable to create a cluster across different availability zones, but there should be a cluster per zone, synchronized as we will see in the *Creating a geographical cluster replication* recipe.

Whether it is a simple, mirrored queue or a geographically replicated cluster, clients can assume to have successfully delivered and secured their messages to the mirrored queues only after they receive the corresponding notifications back (with ack messages using the publisher-confirms RabbitMQ extensions, or by using transactions that are of AMQP standard).

> In case you are dealing with a geographically replicated cluster, after a client publishes a message, RabbitMQ sends the ack message back to the client after the corresponding message has been replicated across the nodes, mirroring the queues in the local cluster. The geographical replication of the message is asynchronous instead. The message published on the local cluster is already to be considered secured and usually, the asynchronous copy to remote clusters is tolerable for disaster-recovery situations.

In the *Client high availability* recipe, we will see how to correctly implement a client that will be able to deal with the conditions that are common when connecting to high-availability RabbitMQ clusters.

Mirroring queues

RabbitMQ clusters don't mirror queues by default. The queues are stored in the broker nodes connected to the clients that created them. Whenever such a node fails, all the queues and the messages stored within it aren't available.

If you have defined the queues as durable and the messages as persistent, it's possible to restore the node without losing data but this is not sufficient.

In fact, designing a highly available application can't be acceptable. There are many cases where the application must be able to survive the death of one component without interruption.

The **ha-policies** help to solve this problem. In this recipe, we will show you how to mirror a queue across all the nodes in the cluster.

Getting ready

You need a RabbitMQ cluster with three nodes.

How to do it...

In order to configure a mirror queue, there are two ways that can be used; that is, it can be configured using `rabbitmqctl` or using the web management plugin (or its API). We will show both the ways in the following steps:

1. First, you need a RabbitMQ cluster, as seen in the *Creating a simple cluster* recipe in *Chapter 6, Developing Scalable Applications*.

2. Configure the policies via web management by navigating to **Admin | Policies | Add / update a policy**.

3. Enter `mirror-all` to the **Name** field, `^mirr\.` to the **Pattern** field, and `ha-mode` and `all` to the **Definition** field.

4. Click on **Add policy**, as shown in the following screenshot:

5. Alternatively, issue the following command via `rabbitmqctl`:

   ```
   rabbitmqctl set_policy ha-all "^mirr\." '{"ha-mode":"all"}'
   ```

6. Create a queue named `mirr.q_connection_1_1`. (It's important that the prefix is `mirr.`; you are free to call the queue as you prefer.)

How it works...

After creating the cluster (refer to step 1), you need to define the mirror queues' behavior using `ha-policies` (refer to step 2).

 The RabbitMQ policies are key-value pairs that can be used to describe the behavior of the federation plugin, mirrored queues, alternate exchanges, dead lettering, per queue TTLs, and maximum queue length.

The `mirror-all` parameter is the policy name, and the `^mirr\.` string is the regular expression pattern. We are instructing RabbitMQ to mirror all the queues that have names starting with `mirr.`.

 Policies are defined using regular expressions, which allow you to create flexible and complex mirror behaviors for your policies.

The last parameter is the mirroring mode. With `ha-mode:all`, the queues are mirrored across all the nodes in the cluster using a master slave pattern (the slaves can be more than one). Whenever a queue matches the given pattern is created on one node, it is replicated on all the others, and whenever a client starts inserting messages in it, they are replicated across all the slaves.

 It's best practice to always connect the clients to the node hosting the master queue and use the other ones only when it fails.

For example, if we have three nodes, namely `rabbit@rabbitmqc1`, `rabbit@rabbitmqc2`, and `rabbit@rabbitmqc3`, and we create a queue with the prefix `mirr.` on the `rabbitmqc1` node (the master for this queue), the queue will be replicated on the other two nodes (the slaves) as well. In order to check the queue's status, open the web management console and click on the **Overview** tab of the queue, as shown in the following screenshot:

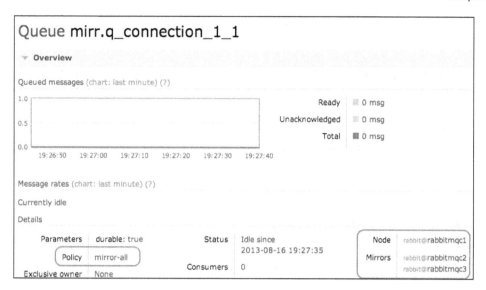

If the `rabbitmqc1` node is down or unavailable, a new master will be promoted for the queue. For example, if we shut it down by calling `rabbitmqctl stop_app`, the `rabbit@rabbitmqc3` node is promoted as the master for the `mirr.q_connection_1_1` queue, as shown in the following screenshot:

When we restart `rabbit@rabbitmqc1`, the queue is mirrored again. Thanks to the defined policy, it is configured as a slave and not promoted to a master as it was originally:

Node	rabbit@rabbitmqc3
Mirrors	rabbit@rabbitmqc2
	rabbit@rabbitmqc1

The cluster is fully mirrored again.

There's more...

In this recipe we have seen how to quickly set up mirrored queues, but it's not enough to build a fully and highly available solution based on RabbitMQ. In fact, the RabbitMQ mirroring configuration lets the broker keep copies of the messages to avoid losing any message. But, it's important to properly handle the connections and the messages on the client side in order to take advantage of this.

In the next recipes, we will learn about the mirror queues in detail, tackling all these aspects.

See also

At `http://www.rabbitmq.com/ha.html`, you can find all the details on RabbitMQ high availability.

For more details on policies, refer to `http://www.rabbitmq.com/parameters.html#policies`.

Synchronizing queues

As we have seen in the *Mirroring queues* recipe, when a mirror is configured, the messages are copied across the cluster.

However, a new node can be added to the cluster at any time and can start to host mirrored queues that already contain messages. How does the cluster behave with respect to the stored messages?

Let's suppose that we have a typical scenario with a standalone node that has some messages stored in one of its queues as follows:

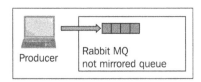

Now, if we add a node to the cluster and properly configure the ha-policies, the queue of the first node gets mirrored and subsequent messages start to get replicated on the newly added node, seen as follows:

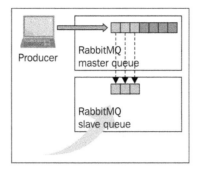

It's very important to note that messages that are already in the master queue at the time of the addition of the second node don't get replicated by default. If the master dies at this time, these messages are lost. However, in a typical case with "live" queues, as soon as the consumer drains the single copy messages, the configuration will become fully replicated without any supplementary effort, as shown in the following screenshot:

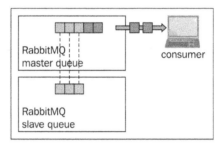

Alternatively, in the case of a queue not being drained, it's important to explicitly synchronize the queues to the mirror in order to have improved reliability. (After all, that's why we are adding a mirror.)

This is not the default behavior, because the synchronization task can have an important impact on the performance of the broker. When starting a synchronization, the queues are stuck until this process is done. In this recipe we will show you how to check the replication status and synchronize the queues.

Getting ready

You need a RabbitMQ cluster with mirror policies (refer to the *Mirroring queues* recipe).

How to do it...

In order to see the behavior of unsynchronized queues, we will simulate a node failure situation manually using the following steps:

1. Configure the mirroring for queues prefixed by `mirr.`, as seen in the *Mirroring queues* recipe (we call the `rabbit@rabbitmqc1` and `rabbit@rabbitmqc2` nodes).

2. Create a queue named `mirr.q_connection_1_1`.

3. Check the queue's status from the web console; you should have the following screenshot:

4. You can also check the queue using `rabbitmqctl list_queues name policy slave_pids`. The result should be as follows:

    `mirr.q_connection_1_1   ha-all all`

    `<rabbit@rabbitmqc1.2.2844.1>`

    `[<rabbit@rabbitmqc2.2.3363.1>]`

    `running`

5. Shut down the `rabbit@rabbitmqc2` node using `rabbitmqctl stop_app` (actually it doesn't matter which node).

6. Publish non-persistent messages to the queue using the `rabbit@rabbitmqc1` node.

7. Restart the application on the `rabbit@rabbitmqc2` node using `rabbitmqctl start_app`. Then, check the queue as shown in the following screenshot:

8. Synchronize the queue using the **Syncronise** button or using `rabbitmqctl sync_queue mirr.q_connection_1_1`:

```
root@rabbitmqc2:~# rabbitmqctl sync_queue mirr.q_connection_1_1
Synchronising queue 'mirr.q_connection_1_1' in vhost '/' ...
...done.
```

How it works...

After creating a stable situation by following steps 1 to 4, we simulate a node failure by stopping the `rabbit@rabbitmqc2` node. So when we publish a message to the queue using the `rabbit@rabbitmqc1` node, the messages aren't mirrored, because we have only one node up and running.

When the `rabbit@rabbitmqc2` node gets running again (step 7), its queue is unsynchronized as we have seen in the *Introduction* section already.

You can synchronize the queue using the web management plugin or `rabbitmqctl`. In this way, your messages are replicated across the cluster.

There's more...

We have seen how to synchronize a queue manually, but there is also an automatic way to do this. Simply add `ha-sync-mode` and `automatic` when you configure `ha-policies`, as in the following screenshot:

If you simulate a failure situation with the two policies, you will have the following screenshot:

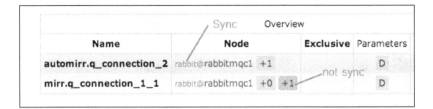

See also

You can see the *Synchronising queues* section at `http://www.rabbitmq.com/ha.html` for more details.

Optimizing mirror policies

In the *Mirroring queues* recipe, we have seen how to mirror a queue across all the nodes in the cluster. Replicating messages on more than two nodes improves the system availability minimally, but if the cluster grows because of the higher load, it will negatively impact the application's performance. In this recipe we'll show you how to distribute each queue to a pair of nodes, in order to have two copies for each queue as follows:

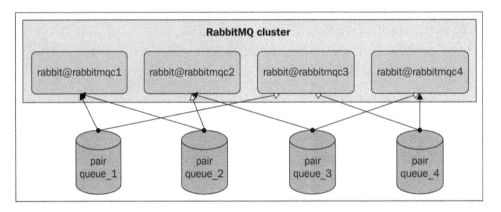

In this way, each queue has one master and one slave.

Getting ready

You need a RabbitMQ cluster with mirror policies (refer to the *Mirroring queues* recipe).

How to do it...

In this recipe we have used four machines; let's see the steps in detail:

1. Create a RabbitMQ cluster with four nodes named `rabbit@rabbitmqc1`, `rabbit@rabbitmqc2`, `rabbit@rabbitmqc3`, and `rabbit@rabbitmqc4`. The following screenshot shows our cluster in action:

Name	File descriptors (?)	Socket descriptors (?)	Erlang processes	Memory	Disk space
rabbit@rabbitmqc1	24 1024 available	1 829 available	198 1048576 available	32.6MB 198.8MB high watermark	17.7GB 953.7MB low watermark
rabbit@rabbitmqc2	28 1024 available	1 829 available	201 1048576 available	32.0MB 198.8MB high watermark	17.7GB 953.7MB low watermark
rabbit@rabbitmqc3	21 1024 available	1 829 available	195 1048576 available	30.7MB 198.8MB high watermark	17.7GB 953.7MB low watermark
rabbit@rabbitmqc4	21 1024 available	1 829 available	195 1048576 available	31.0MB 198.8MB high watermark	17.7GB 953.7MB low watermark

2. Create an `ha-policy` using the following code:

```
"mirr-pair" as name,
"^pair\."as pattern
   "ha-mode":"exactly"
   "ha-params":   2
   "ha-sync-mode":"automatic" as parameters
```

Refer to the following screenshot:

3. Create a queue and name it with the prefix `pair` as you wish. You can directly use the web management console:

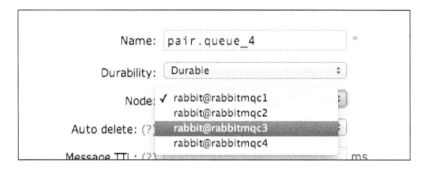

How it works...

In step 2, we configured `ha-mode` with `exactly`. This parameter needs the `ha-params` parameter, which indicates the number of nodes that will mirror the queues, which in our case are two nodes. A new queue with the prefix `pair` (refer to step 3) will be created as a master on the node where we have performed the operation, and it will get just one slave on another node.

When dealing with real applications, that is, not connecting to the management console, we create the queues from the application code by invoking `channel.queueDeclare()`. In this case, the node to which we are connected will create the master queue.

In order to evenly distribute the queues to the cluster, it's important to connect to all the nodes using a load balancer or letting the client connect in round-robin to all the available nodes of the cluster, as we will see in the *Client high availability* recipe.

The HA/Mirroring feature adds overhead and may decrease performance.

There's more...

The `exactly` parameter chooses a node slave automatically from the cluster. However, you can use more parameter nodes to choose where the mirrored queues will be exactly placed, as you can see in the following screenshot:

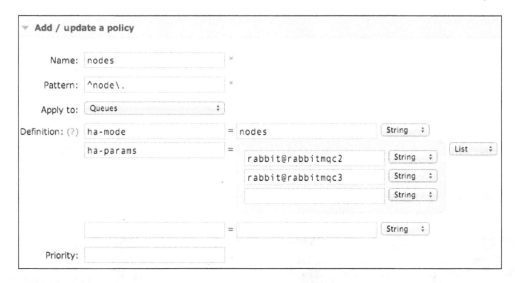

See also

Read the *Configuring Mirroring* section at `http://www.rabbitmq.com/ha.html`.

Distributing messages between a couple of brokers

The `ha-mirror` plugin requires a cluster. As we have seen in *Chapter 6, Developing Scalable Applications*, it does not tolerate network partitions well.

In order to replicate messages across the WAN, you can use the federation plugin. This plugin does not require a cluster, so you can federate more RabbitMQ instances over a WAN even with different Erlang versions.

Getting ready

You need two or more RabbitMQ nodes. In this example, we have used two Linux machines whose RabbitMQ node names are `rabbit@rabbitmqc1` and `rabbit@rabbitmqc2`.

How to do it...

The federation plugin must be enabled; it's disabled by default. For both the machines, perform the following steps:

1. Enable the plugin using the following command:

    ```
    rabbitmq-plugins enable rabbitmq_federation
    ```

2. Enable the plugin for the web management console using the following command:

    ```
    rabbitmq-plugins enable rabbitmq_federation_management
    ```

3. Restart RabbitMQ and check the plugin using `rabbitmqctlstatus`, as you can see in the following screenshot:

```
root@rabbitmqc1:~# rabbitmqctl status
Status of node rabbit@rabbitmqc1 ...
[{pid,2791},
 {running_applications,
     [{rabbitmq_federation_management,"RabbitMQ Federation Management",
          "3.1.4"},
      {rabbitmq_management,"RabbitMQ Management Console","3.1.4"},
      {rabbitmq_web_dispatch,"RabbitMQ Web Dispatcher","3.1.4"},
      {webmachine,"webmachine","1.10.3-rmq3.1.4-gite9359c7"},
      {mochiweb,"MochiMedia Web Server","2.7.0-rmq3.1.4-git680dba8"},
      {rabbitmq_management_agent,"RabbitMQ Management Agent","3.1.4"},
      {rabbitmq_federation,"RabbitMQ Federation","3.1.4"},
```

For `rabbitmqc1` machine, perform the following steps:

4. It's also possible to check the status on the web management. On selecting the **Admin** tab, you can see the **Federation Status** and **Federation Upstreams** entries on the right, as shown in the following screenshot:

5. Configure the federation upstreams. Open the web management console by navigating to **Admin | Federation Upstreams | Add a new upstream**, and then fill the following fields:

   ❑ **Name**: first_upstream

   ❑ **URI**: amqp://rabbitmqc2

Add a new upstream	
Name:	first_upstream
URI: (?)	amqp://rabbitmqc2
Expires: (?)	ms
Message TTL: (?)	ms
Max hops: (?)	
Prefetch count: (?)	

6. For the rabbitmqc1 machine, configure the federation policy. Open the web management console and navigate to **Admin | Policies | Add / update a policy**, and fill the following fields:

   ❑ **Name**: fed_policy

   ❑ **Pattern**: ^fed\.

   ❑ **Definition**: federation-upstream-set:all

7. For the rabbitmqc1 machine, add a new exchange with the prefix fed., for example fed.myfanoutexchange.

8. For the rabbitmqc1 machine, check the upstream status from the web management console by navigating to **Admin | Federation Status** and if everything is fine, you should have the following screenshot:

Running Links

ᴐnnection	URI	Exchange	State
_upstream	amqp://rabbitmqc2	**fed.myfanoutexchange**	running

How it works...

The RabbitMQ federation plugin must be enabled by following steps 1 and 2. You can either use the command-line tool or the web management console to verify its status.

In order to put the federation to work, you have to define an upstream link (refer to step 4). You can perform the definition on the downstream node, `rabbitmqc1`, and specify which is the upstream node, `rabbitmqc2`.

 The federation plugin doesn't require a cluster, and the URI configuration can use an IP address, for example, `amqp://192.168.0.23`. There's no need to have short hostnames here.

The messages published to the federated exchanges on the upstream node will propagate downstream on the corresponding exchanges.

 The federation is strictly one-way. The messages published to the downstream node won't get copied to the upstream.

On the `rabbitmqc1` node when you create an exchange with the prefix `fed.`, for example `fed.myfanoutexchange`, the exchange will also be created to the `rabbitmqc2` broker.

The messages are replicated in an asynchronous way. Differently from queue mirroring, you have no high-availability guarantees.

There's more...

The federation plugin is very easy to configure, but it contains a lot of parameters to manage different situations. Here we have seen the base configuration. Read the documentation at `http://www.rabbitmq.com/federation.html` for more information.

Creating a geographical cluster replication

Besides `ha-mirror` and federation plugins, there is a shovel plugin to increase the application reliably. The shovel is a queue-to-exchange pumper. It's a sort of RabbitMQ client integrated in the broker as a plugin that can consume messages from one or more queues and redirect them to other ones either locally or on remote brokers.

As a client, it's ready for WAN connections and can tolerate network interruptions.

In this recipe, we are going to show you how to use the plugin with an easy configuration. We will pump messages between two queues on different brokers with a WAN connection in between, which can be seen as follows:

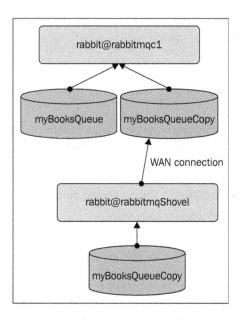

Getting ready

You need two RabbitMQ instances; name them `rabbitmq@rabbitmqc1` and `rabbitmq@rabbitmqShovel`.

How to do it...

You need to install the shovel plugin in one of the brokers (`rabbitmq@rabbitmqShovel`) by performing the following steps:

1. Enable the plugin using `rabbitmq-plugins enable rabbitmq_shovel`.

2. Enable the plugin for web management using `rabbitmq-plugins enable rabbitmq_shovel_management`.

3. Edit, or if it doesn't exist, create the `rabbitmq.config` file and add the shovel configuration as follows:

```
[{rabbitmq_shovel,
  [{shovels,
    [{my_books_shovel,
      [
        {sources,
        [{broker, "amqp://yourrabbitmqip"}]},
        {destinations, [ {broker, "amqp://"}]}
        , {queue, <<"myBooksQueueCopy">>}
```

```
                  , {prefetch_count, 10}
                  , {reconnect_delay, 5}
              ]}
           ]}
        ]}
     ].
```

You can use the file from the book example's sources located at `Chapter07/Recipe05/simple_shovel_rabbitmq.config`.

4. Restart `broker`.

5. Create the queues as shown in the preceding diagram.

6. Check the shovel from web management.

How it works...

In this example, the plugin consumes messages from the `myBooksQueueCopy` queue located at `rabbitmq@rabbitmqc1` and publishes them to the `myBooksQueueCopy` queue located at `rabbitmq@rabbitmqShovel`. The two brokers can be geographically separated since the shovel plugin is actually an embedded RabbitMQ client consuming the messages from one peer and republishing them to the other one.

The shovel starts at the RabbitMQ startup and in this example, it will start polling the source queue contents, continuing to poll even if the queue itself is not defined.

The source and destination parameters are mandatory.

 The plugin uses the localhost as the broker IP if it is not specified with a URI (`amqp://`).

We need to create a copy of the queue because the messages are consumed by the plugin. So the `myBooksQueue` queue is consumed by the application's consumer, and `myBooksQueueCopy` is consumed by the shovel.

There's more...

The shovel is a low-level client and usually doesn't have an impact on the system's performance. It has other parameters that you can see at `http://www.rabbitmq.com/shovel.html`.

In this recipe we have realized a simple message pump. In the next recipe, we will see how to bind a shovel dynamically.

Filtering and forwarding messages

In this recipe we are going to implement a selective message forwarding. We will let the shovel plugin forward the subsets of messages to different destinations. A possible use case of this example is a company with three different sites having the following responsibilities:

- One that takes book orders
- One that needs only the orders with `london` as the routing key
- One that needs only the orders with `rome` as the routing key

The aim is to forward messages selectively by adding or removing shovels without touching the source broker as follows:

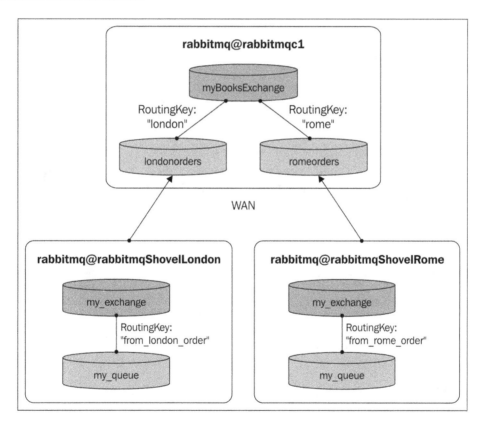

Getting ready

You need three brokers; we will name them `rabbitmq@rabbitmqc1`, `rabbitmq@rabbitmqShovelLondon`, and `rabbitmq@rabbitmqShovelRome`.

How to do it...

In this example the `rabbitmq@rabbitmqc1` broker is the node that the shovels will connect to and the plugin is not necessary, but we need to enable it for the others:

1. Enable the shovel plugin on the `rabbitmq@rabbitmqShovelLondon` and `rabbitmq@rabbitmqShovelRome` nodes as seen in the *Creating a geographical cluster replication* recipe.

2. Create one shovel script for the `rabbitmq@rabbitmqShovelLondon` node:

   ```
   {sources, [ {broker, "amqp://rabbitmqc1IP"},
   {declarations, [ 'queue.declare'
   {routing_key, <<"london">>}
   . . . .
   , {destinations, [ {broker, "amqp://"}]}
   , {prefetch_count, 10}
   ,{publish_fields, [ {exchange, <<"my_exchange">>}, {routing_key,
   <<"from_london_order">>} ]}
   . . .
   ```

   You can find the full configuration at `Chapter07/Recipe06/london_shovel_rabbitmq.config`.

3. Create one shovel script for the `rabbitmq@rabbitmqShovelRome` node:

   ```
   {sources, [ {broker, "amqp://rabbitmqc1IP"},
   {declarations, [ 'queue.declare'
   . . .
   {routing_key, <<"rome">>}
   . . .
   , {destinations, [ {broker, "amqp://"}]}
   , {prefetch_count, 10}
   ,{publish_fields, [ {exchange, <<"my_exchange">>}, {routing_key,
   <<"from_rome_order">>} ]}
   . . .
   ```

   You can find the full configuration at `Chapter07/Recipe06/rome_shovel_rabbitmq.config`.

4. Restart the brokers.

5. Publish two messages to `myBooksExchange`, one using the `london` routing key and the other using the `rome` routing key.

How it works...

The scripts defined at steps 2 and 3 contain the declarations for the exchanges and queues for each broker. So as soon as you run the declarations, the shovel plugins will declare and create two queues in the remote node (by default), as shown in the following screenshot:

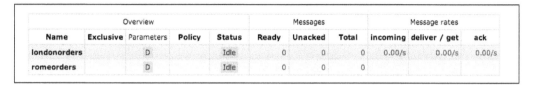

Name	Exclusive	Overview Parameters	Policy	Status	Messages Ready	Unacked	Total	Message rates incoming	deliver / get	ack
londonorders		D		Idle	0	0	0	0.00/s	0.00/s	0.00/s
romeorders		D		Idle	0	0	0			

The `londonorders` queue has been declared by `rabbitmq@rabbitmqShovelLondon`, and `romeorders` has been declared by `rabbitmq@rabbitmqShovelRome`.

The scripts also declare `myBooksExchange` on the same node and bind it to the two queues using `london` and `rome` as the routing key respectively, as you can see in the following screenshot:

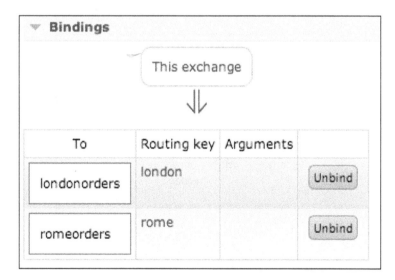

In this way, when you publish a message to `myBooksExchange` with the `rome` routing key, the message gets routed to the local `romeorders` queue, and then is immediately consumed by the shovel running on `noderabbitmq@rabbitmqShovelRome`.

Once it reaches the node, the shovel publishes it to the local topic exchange `my_exchange` and it finally gets routed to the queue `my_queue` with a new routing key named `from_rome_order`, independent from the original one as you can see in the following screenshot:

> ## Publish message
>
> Routing key: `rome`
>
> Delivery mode: `1 – Non-persistent`
>
> Headers: (?)
>
> Properties: (?)
>
> Payload:
> ```
> Book published!
> Exchange: myBooksExchange
> Node: rabbitmq@rabbitmqc1
> RoutingKey: rome
> ```
>
> ---
>
> Message 1   rabbitmqShovelRome
>
> The server reported 0 messages remaining.
>
> | | |
> |---|---|
> | Exchange | my_exchange |
> | Routing Key | from_rome_order |
> | Redelivered | ○ |
> | Properties | delivery_mode: 1 |
> | | headers: |
> | Payload | Book published! |
> | 89 bytes | |
> | Encoding: string | Exchange: myBooksExchange |
> | | Node: rabbitmq@rabbitmqc1 |
> | | RoutingKey: rome |

The same happens with `london` on `rabbitmq@rabbitmqShovelLondon`.

Even if a downstream broker loses a connection, no messages will be lost, because the messages are stored on the persistent queues of the `rabbitmq@rabbitmqc1` node.

There's more...

The shovel plugin is a very powerful tool, which can be used in various contexts; for example, you can use it on a single broker to create an asynchronous queue mirror.

See also

Refer to `http://www.rabbitmq.com/shovel.html#configuration` for more details on each single instruction used in this example.

Combining high-availability technologies together

RabbitMQ has three different ways to distribute messages among brokers:

- ▸ Cluster queue mirroring
- ▸ Federation
- ▸ Shovel

In this recipe we will show you how to combine cluster, high-availability queue mirroring, and shovel to transfer messages across the WAN from an `ha-cluster` to a single RabbitMQ node, as shown in the following diagram:

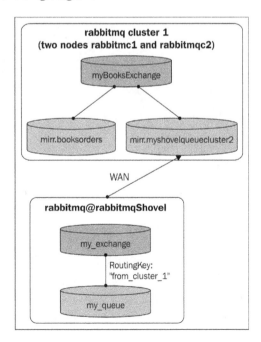

The aim is to implement a shovel capable of draining messages from a cluster using its high-availability properties.

Getting ready

You need three RabbitMQ nodes.

How to do it...

You should have two nodes on the same LAN, and the third node eventually should be outside the network. We will name the nodes `rabbit@rabbitmqc1`, `rabbit@rabbitmqc2`, and `rabbit@rabbitmqShovel` as follows:

1. Create a cluster using `rabbit@rabbitmqc1` and `rabbit@rabbitmqc2`, as seen in the *Creating a Simple Cluster* recipe in *Chapter 6, Developing Scalable Applications*. We will refer to the cluster as `Cluster1`.

2. In `Cluster1`, create a topic exchange on the cluster called `myBooksExchange`.

3. In `Cluster1`, create an `ha-queue` as seen in *Mirroring queues* and name it `mirr.orders`. Bind the queue to `myBooksExchange` using # as the routing key.

4. In `rabbit@rabbitmqShovel`, enable the shovel plugin as seen in the *Creating a geographical cluster replication* recipe.

5. In `rabbit@rabbitmqShovel`, create or edit `rabbitmq.config` to add the shovel configuration as follows:

You can find the file in the code bundle of the book at `Chapter07/Recipe07/rabbitmq.config` and copy it directly to the configuration folder.

6. In `rabbit@rabbitmqShovel`, restart the broker.

How it works...

In the steps 1 and 2, we created a cluster and a couple of mirrored queues. The shovel configuration (refer to step 5) creates a queue named `mirr.myshovelqueuecluster1`. The queue is mirrored on the two RabbitMQ nodes. You should have at least two mirrored queues in the cluster:

	Overview	
Name	**Node**	**Exc**
mirr.myshovelqueuecluster1	rabbit@rabbitmqc1 +1	
mirr.orders	rabbit@rabbitmqc1 +1	

The `brokers` parameter in the shovel configuration contains both the node cluster addresses. If one node fails, the shovel will connect to the other node where the `mirr.myshovelqueuecluster1` queue is already present, since it is mirrored. This configuration can lose data only if both the nodes of `Cluster1` are down; in other situations, you won't lose any data.

There's more...

On using the shovel plugin, it is possible to create some odd topologies where messages are being replicated without pace, in particular, putting two or more shovels in a loop. So if you want to create a bidirectional replication, you should use the federation plugin in order to avoid an infinite loop. In this case, you have the `max-hops` parameter for this purpose.

See also

Read the *Summary* section at `http://www.rabbitmq.com/distributed.html` to view the differences between cluster, federation, and shovel.

Client high availability

Even if the RabbitMQ broker provides many high-availability server-side options, they are pretty useless if the connecting clients do not implement some measures in spite of extending the high availability to the clients too:

- Clients must check that the produced messages are successfully transferred to RabbitMQ and handle error conditions, for example, republishing them if needed.
- Clients consuming messages must check that they are not duplicated; given the possibility of retransmitting messages, it is possible to actually get duplicated messages on the consumer side too. This activity is commonly called **deduplication**.
- Clients must check that the RabbitMQ node to which they are connected is healthy. In particular, a consumer waiting for messages could not realize that it is not receiving messages because the server is stuck.
- Clients should try to connect to any available cluster node, both for maximum reliability and for uniform resource distribution.

These mechanisms are not built into the client libraries but must be implemented following some specific guidelines (`http://www.rabbitmq.com/reliability.html` and `http://www.rabbitmq.com/ha.html#behaviour`) that we are going to represent with this recipe.

As always, increased reliability comes at the price of performance: the cluster configuration, the queue mirroring, the checks needed for both producing and consuming messages, increase the per-message latency and lower the maximum message rate.

Mostly, nothing can be done to lower the latency in these conditions, but it is possible to handle higher message rates by growing the cluster. Here, the scalability of RabbitMQ plays its role.

We will present how to develop clients that reliably produce and consume messages to/from a mirrored queue on a RabbitMQ cluster.

Getting ready

To run this recipe you need the following tools:

- A RabbitMQ cluster with at least two nodes
- The ha-configuration shown in the *Mirroring queues* recipe; mirroring queues matching the regular expression `^mirr\.`
- The RabbitMQ Java Client API

How to do it...

This example is composed of two Java programs, `ProducerMain.java` and `ConsumerMain.java`, which you can find in the code bundle of the book under the path `Chapter09/Recipe08` using the `ReliableProducer.java` and `ReliableConsumer.java` files respectively, the core classes of the example. The rest of the sources of the project are mostly shared with the full view of the main components used in the example summarized in the UML class diagram, as shown in the following screenshot:

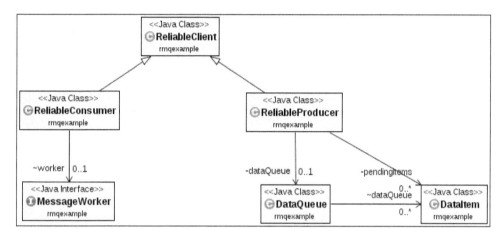

The preceding screenshot shows the steps needed to develop such reliable clients.

Let's start with the common steps for both producers and consumers:

1. Write a general method that will keep trying to open a connection until it succeeds (open the `ReliableClient.java` file):

```
protected void waitForConnection() throws
  InterruptedException {
  while (true) {
    ConnectionFactory factory = new ConnectionFactory();
    ArrayList<Address> addresses = new
      ArrayList<Address>();
    for (int i = 0; i<Constants.hosts.length; ++i) {
      addresses.add(new Address(Constants.hosts[i],
        Constants.port));
    }

    // randomize the order used to try the servers:
    // distribute their usage
```

```
      Collections.shuffle(addresses);
      Address[] addrArr=new Address[Constants.hosts.length];
      addresses.toArray(addrArr);
      try {
        connection = factory.newConnection(addrArr);
        channel = connection.createChannel();
        channel.exchangeDeclare(Constants.exchange, "direct",
          false);
        channel.queueDeclare(Constants.queue,
          Constants.durableQueue, Constants.exclusiveQueue,
            Constants.autodeleteQueue, null);
        channel.queueBind(Constants.queue,
          Constants.exchange,Constants.routingKey);
        return;
      } catch (Exception e) {
        e.printStackTrace();
        disconnect();
        Thread.sleep(1000);
      }
    }
  }
```

2. Write a `disconnect` method (open the `ReliableClient.java` file):

```
protected void disconnect() {
  try {
    if (channel != null && channel.isOpen()) {
      channel.close();
      channel = null;
    }
    if (connection != null && connection.isOpen()) {
      connection.close();
      connection = null;
    }
  } catch (IOException e) {
    // just ignore
    e.printStackTrace();
  }
}
```

Then let's see how to put a reliable producer to work (open the
`ReliableProducer.java` file):

3. Extend the `ReliableClient` class and override the `waitForConnection()` method as follows:

```java
public class ReliableProducer extends ReliableClient {
...
  @Override
  protected void waitForConnection() throws
    InterruptedException {
    super.waitForConnection();
    try {
      channel.confirmSelect();
    } catch (IOException e) {
      e.printStackTrace();
    }
    channel.addConfirmListener(new ConfirmListener() {
      @Override
      public void handleAck(long deliveryTag, boolean
        multiple) throws IOException {
        if (multiple) {
          ReliableProducer.this.
            removeItemsUpto(deliveryTag);
        } else {
          ReliableProducer.this.removeItem(deliveryTag);
        }
      }

      @Override
      public void handleNack(long deliveryTag, boolean
        multiple) throws IOException {
        if (multiple) {
          ReliableProducer.this.
          requeueItemsUpto(deliveryTag);
        } else {
          ReliableProducer.this.requeueItem(deliveryTag);
        }
      }

    });
  }
```

4. Write a method that lets the `ReliableProducer` class actually publish messages from its internal `dataQueue`:

```
protected void publishFromLocalQueue() throws
  InterruptedException {
  try {
    for (;;) {
      synchronized (dataQueue) {
        if (dataQueue.isEmpty()) {
          dataQueue.wait(1000);
          // if the queue stays empty for more than
          // one second, disconnect and
          // wait offline
          if (dataQueue.isEmpty()) {
            System.out.println("disconnected for
              inactivity");
            disconnect();
            dataQueue.wait();
            waitForConnection();
          }
        }
      }
      DataItem item = dataQueue.peek();
      BasicProperties messageProperties = new
        BasicProperties.Builder()
      .messageId(Long.toString(item.getId()))
      .deliveryMode(2)
      .build();
      long deliveryTag = channel.getNextPublishSeqNo();
      channel.basicPublish("", Constants.queue,
        messageProperties, item.getData().getBytes());
      // only after successfully publishing,
      // move the item to the
      // container of pending items.
      // They will be removed from it only
      // upon the
      // reception of the confirms from the broker.
      synchronized (pendingItems) {
        pendingItems.put(deliveryTag, item);
      }
      dataQueue.remove();
      if (Thread.interrupted()) {
        throw new InterruptedException();
      }
```

```
      }
    } catch (IOException e) {
      // do nothing: the connection will be closed
      // and then retried
    }
  }
```

5. Write the main loop method invoked by the `main()` method of `ProducerMain.java`. This one starts a background thread, which waits for a connection and asynchronously publishes the messages in the local queue:

```
public void startAsynchronousPublisher() {
  exService = Executors.newSingleThreadExecutor();
  exService.execute(new Runnable() {
    @Override
    public void run() {
      try {
        for (;;) {
          waitForConnection();
          publishFromLocalQueue();
          disconnect();
        }
      } catch (InterruptedException ex) {
        // disconnect and exit
        disconnect();
      }
    }
  });
}
```

6. Write the method of `ReliableProducer` invoked from `ProducerMain.java`, which temporarily stores the messages in the local `dataQueue`:

```
public void send(String data) {
  synchronized (dataQueue) {
    dataQueue.add(data);
    dataQueue.notify();
  }
}
```

7. Here, `dataQueue` is an instance of a thread-safe `DataQueue` class that contains a unique index too:

```
public class DataQueue {
  ...
```

```
         public synchronized long add(String data) {
            ++lastID;
            dataQueue.add(new DataItem(data,lastID));
            returnlastID;
         }
      ...
      }
```

Let's now see the steps needed in the `ReliableConsumer` class:

8. In this case too we override the connection method named `ReliableClient.WaitForConnection()`:

```
public class ReliableConsumer extends ReliableClient {
...
   @Override
   protected void waitForConnection() throws
      InterruptedException {
      super.waitForConnection();
      try {
        channel.basicConsume(Constants.queue, false, new
          Consumer() {

          @Override
          public void handleCancel(String consumerTag) throws
            IOException {
            System.out.println("got handleCancel signal");
          }

          @Override
          public void handleCancelOk(String consumerTag) {
            System.out.println("got handleCancelOk signal");
          }

          @Override
          public void handleConsumeOk(String consumerTag) {
            System.out.println("got handleConsumeOK signal");
          }

          @Override
          public void handleDelivery(String consumerTag,
            Envelope envelope,BasicProperties properties,
              byte[] body) throws IOException {
```

```
        long messageId =
          Long.parseLong(properties.getMessageId());
      if (worker != null) {
        // if the message is not a re-delivery,
        // sure it is not aretransmission
        if (!envelope.isRedeliver() ||
          toBeWorked(messageId)) {
          try {
            worker.handle(new String(body));
            // the message is ack'ed just after it has
            // been
            // secured (handled, stored in database...)
            setAsWorked(messageId);
            channel.basicAck(envelope.getDeliveryTag(),
              false);
          } catch (WorkerException e) {
          // the message worker has reported
          // an exception,
          // so the message
          // cannot be considered to be handled
          // properly,so requeue it
          channel.basicReject
            (envelope.getDeliveryTag(), true);
        }
      }
    }
  }

  @Override
  public void handleRecoverOk(String consumerTag) {
    System.out.println("got recoverOK signal");
  }

  @Override
  public void handleShutdownSignal(String consumerTag,
    ShutdownSignalException cause) {
    System.out.println("got shutdown signal");
  }
});
} catch (IOException e) {
e.printStackTrace();
}
}
```

9. Write the method that starts the asynchronous consumer of `ReliableConsumer` in a background thread:

```java
public void StartAsynchronousConsumer() {
  exService = Executors.newSingleThreadExecutor();
  exService.execute(new Runnable() {

    @Override
    public void run() {
      try {
        for (;;) {
          waitForConnection();
          synchronized (this) {
            this.wait(5000);
          }
          disconnect();
        }
      } catch (InterruptedException ex) {
        disconnect();
      }
    }
  });
}
```

10. Let the `ReliableConsumer` class allow the setting up of a callback obeying the following interface:

```java
public interface MessageWorker {
  public void handle(String message) throws
    WorkerException;
}
```

11. The code must be passed in `ConsumerMain.java` and will be called on the reception of each message:

```java
reliableConsumer.setWorker(new MessageWorker() {
  @Override
  public void handle(String message) throws WorkerException
  {
    System.out.println("received: " + message);
    ++count;
  }
});
```

How it works...

In order to connect reliably to a cluster, it is important to have the possibility to connect to more than one node. This can be done explicitly, as shown in the example—and in this case, it's the client that loops over the different nodes—or by using a load balancer.

 Depending on the case, it can be useful to randomly connect to any node of the cluster (this kind of approach is named "**active-active**" configuration: both the master and the slave are active, as shown in this case), or preferably to the master, or to the slave when the master is not available (that is called **active-passive** configuration).

Then at the producer side whenever the real application sends a message, the message is not sent actually, but put in a staging queue. Until the message is there, the application must assume that the message has not been received by the broker for sure.

As soon as the background loop (refer to step 4) successfully sends an item, it is pulled from the queue and stored in the `pendingItems` hash map. Even now the application cannot be sure that the message has been successfully stored on all the nodes of the mirrored queue. If a node is down, the message will be lost.

 As we have already pointed out previously, even if we are using mirrored queues, we need to ensure that we do not lose messages when publishing and when consuming.

The `ReliableProducer` class has this guarantee only when it receives a confirmation from the broker, that is, it has been checked by appropriately coding `channel.addConfirmListener()` (refer to step 3). At this point, the message is removed from the map using the delivery tag.

Note that in the example there is no mechanism for retransmitting messages that have not been confirmed after a given timeout (or whatever condition). This needs to be added in a real application by using mechanics, depending upon the real use case.

 AMQP 0-9-1 doesn't contain `confirms`. If strict standard adherence is required, you need to use transactions instead, which are less efficient in the order of magnitude since they behave synchronously.

We are half way now; we can be sure that the message is secured on a mirrored queue in an ha-cluster. Wow! But, now we need to guarantee that we do not lose it while consuming messages.

And that's not all. We need to be sure that we do not consume the same message multiple times.

The connection method for the consumer actually starts the background consumer too by invoking `channel.basicConsume()` and properly defining the `handleDelivery()` callback (refer to step 8).

As seen in *Chapter 1, Working with AMQP*, if everything is fine within the callback, the message is received. The user calls back `worker.handle()` that is invoked and the `ack` message is sent back to the broker, which at this point, deletes it since it has been properly consumed.

But if the user callback raises `WorkerException`, the client rejects the message. Again at this point, the example is open; the message in this case is requeued to the same queue (the second parameter of `channel.basicReject()` is set to `true`), but it can also be redirected to a dead-letter queue for example, or rejected and lost depending on the real application.

However, it is possible that the message is consumed and the callback is properly invoked, but the `ack` message never reaches the application due to a network issue or the broker being abruptly shut down. Here is where the mysterious item ID set in the `DataQueue` instance (refer to step 7) comes into picture. It's needed to avoid duplicates at the receiver side. In fact, RabbitMQ will retransmit the message in the next reconnection since it has not been acknowledged.

 Because RabbitMQ can retransmit messages whenever it suspects that we are in a "risky" condition, we need a mechanism that guarantees the avoidance of message duplicates.

In the recipe, we have assumed to have a monotonically increasing value, starting from zero, which marks the needed messages—again, real applications can have different approaches. In our case, we keep a track of the last message received "continuously" with the `lastItem` sequential counter and of the messages received after some "hole" with the `moreReceivedItems` set.

In this recipe we have used a counter for the following three times to guarantee reliability:

▸ The `deliveryTag` parameter (refer to step 4), which is on the producer side, is needed to check for the confirms at the producer side and delete the items from the `pendingItems` hash map

▸ The `deliveryTag` parameter (refer to step 8), which is on the consumer side, is needed to send the proper `ack` message to the RabbitMQ cluster

▸ The `DataQueue` item ID (refer to steps 7 and 8) is needed to avoid duplicates on the consumer side

8
Performance Tuning for RabbitMQ

In this chapter we will cover:

- ▶ Multithreading and queues
- ▶ System tuning
- ▶ Improving bandwidth
- ▶ Using different distribution tools

Introduction

There are no standard RabbitMQ tuning guidelines because different applications should be optimized in different ways.

Very often the application needs to be optimized on the client side:

- ▶ CPU-intensive applications can be optimized by running one thread for each CPU core
- ▶ I/O-intensive applications can be optimized by running many threads per core in order to hide implicit latencies

In both cases messaging is a perfect fit. In order to optimize the network transfer rate, the AMQP standard prescribes that messages are transferred in bunches, and then consumed one by one by the client (refer to *Chapter 1, Working with AMQP*).

RabbitMQ allows multithreaded applications to consume messages efficiently; this is covered in the *Multithreading and queues* recipe.

Another frequent use case is when RabbitMQ is at the foundation of a distributed application serving a large number of clients. In this case, it's more realistic that the bottleneck is the broker and not the client application.

In this case, it's important that our broker has one characteristic, that is, scalability.

When the number of clients outgrow the current maximum capacity, it's sufficient to add one or more nodes to the RabbitMQ cluster to distribute the load and improve the total throughput.

Why to optimize then? Well, to reduce cost. The cost of the hardware, electric power, cooling, or cloud computing resources.

In this chapter, we will talk about the RabbitMQ performance, showing some tips to improve the performance on both the client side and eventually modifying the broker parameters.

Multithreading and queues

Using threads can help the application's performance. In this recipe, we will show how to use connections, channels, and threads. In this example, we use Java, but generally, using threads is a good practice to improve the performance in most of the current technologies.

You can find the source code in the book archive under the path: `Chapter08/Recipe01`.

Getting ready

You need Java 1.7 or higher and Apache maven.

How to do it...

In this example, we have extended the `ReliableClient` Java class (refer to *Chapter 7, Developing High-availability Applications*) to create a producer and a consumer. Let's see the following steps in detail:

1. Create a maven project and add RabbitMQ client dependency.

2. Create a producer class that extends `ReliableClient`.

3. Create a consumer class that extends `ReliableClient`.

4. For both consumer and producer classes create an `ExecutorService` Java class using the following method:

   ```
   ExecutorServiceexService =
       Executors.newFixedThreadPool(threadNumber);
   ```

5. Create as many `Runnable` tasks as the number of threads. The producer is as follows:

```
for (int i = 0; i<threadNumber; i++) {
  exService.execute(new Runnable() {
    @Override
    public void run() {
      try {
        publishMessages();
```

6. Create as many `Runnable` tasks as the number of threads. The consumer is as follows:

```
for (int i = 0; i<threadNumber; i++) {
  exService.execute(new Runnable() {
    @Override
    public void run() {
      final Channel internalChannel;
      try {
        internalChannel = connection.createChannel();
        @Override
        public void handleDelivery(String consumerTag,
          Envelope envelope, BasicProperties properties,
            byte[] body) throws IOException {..}
```

How it works...

The `ReliableClient` class creates a queue called `perf_queue_08/01`, to which are bound one producer and one consumer. Both the producer and the consumer open one connection and create one channel for each thread. The channel can be shared between more threads, but it's better to create one channel per thread to avoid synchronization times and in some cases lock problems.

> The channel cannot always be thread safe. It depends on the implementation, for example, using the .NET client API you should lock `IModel` before using its methods. Read the *IModel should not be shared between threads* section at `https://www.rabbitmq.com/releases/rabbitmq-dotnet-client/v3.1.5/rabbitmq-dotnet-client-3.1.5-user-guide.pdf`.

To start the threads (refer to step 3), we use the Java `ExecutorService` class with `Executors.newFixedThreadPool(..)`. In this way, you can control your number of threads.

 Please find more information about the `ExecutorService` class and Java thread pooling at `http://docs.oracle.com/javase/7/docs/api/java/util/concurrent/ExecutorService.html`.

In this example, you can choose the message size, the running time, and the consumer message count. You can create the `rmqThreadTest.jar` file using the following command:

```
mvn clean compile assembly:single
```

You can now test the producer with the following command:

```
java -cp rmqThreadTest.jar rmqexample.ProducerMain 4 10000 128
```

The first parameter is the number of threads; the second one is the running time in milliseconds; and the last one the buffer size in bytes.

You can test the consumer using the following command:

```
java -cp rmqThreadTest.jar rmqexample.ConsumerMain 4
```

The parameter is the number of threads. You can combine the producer and consumer parameters to test your application and find the better performance for your environment. Open the web management console to check the actual rate, as shown in the following screenshot:

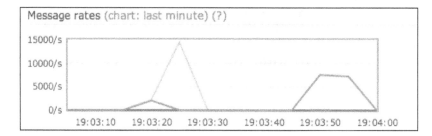

There's more...

The queue is faster when empty and you should design your application, in order to always have a queue as empty as possible. The queue capacity comes in handy every time your application needs to deal with load spikes. By using queues, messages can eventually be buffered and handled without losing any information.

If the consumer is slower than the producer and you have to consume the messages quickly, you can try to add more threads or more consumers.

Since version 3.2.0 RabbitMQ supports the queue federations (`http://www.rabbitmq.com/blog/2013/10/23/federated-queues-in-3-2-0/`), which balance the messages on two or more brokers without having a cluster. Anyway, if you have a previous RabbitMQ version you should split your queues manually, distributing them to more brokers.

The number of threads for the producer and the consumer depends strictly on your application and the deploy environment. Be careful not to open too many threads because you could have the opposite effect.

System tuning

In this recipe, we will appreciate some steps useful to obtain the maximum performance from RabbitMQ. We will cover the following topics:

- The `vm_memory_high_watermark` configuration (`http://www.rabbitmq.com/memory.html`)
- Erlang **High Performance Erlang (HiPE)** (`http://erlang.org/doc/apps/hipe/`)

The `vm_memory_high_watermark` configuration is the maximum percentage of the system memory used to cache messages before they are consumed or cached to the disk.

Before the limit is reached, by default, at fifty percent of `vm_memory_high_watermark`, (or properly setting the `vm_memory_high_watermark_paging_ratio` parameter, set to `0.5` by default), RabbitMQ will start to move messages from memory to on-disk paging space.

If neither this paging mechanism, nor the consumers are able to keep pace with the producers, the limit will be reached, and then RabbitMQ will block the producers.

In some cases, it is possible to enlarge these parameters, in order to avoid starting to page messages to the disk too early. In this recipe, we will see how to do it in conjunction with HiPE. There are two different aspects, but the steps needed to accomplish them are very similar.

You can use the code from the book repository in the directory `Chapter08/Recipe02`.

Getting ready

To try this recipe, you need to start with RabbitMQ and have the management plugin installed. Then, you need Java 1.7 or higher and Apache maven.

How to do it...

In order to obtain the maximum performance from RabbitMQ, you can perform the following steps:

1. Configure the watermark using:

   ```
   rabbitmqctl set_vm_memory_high_watermark 0.6
   ```

   Or directly in the rabbitmq.config file using:

   ```
   [{rabbit, [{vm_memory_high_watermark, 0.6}]}].
   ```

2. Change the Linux `ulimit` parameter modifying the `/etc/default/rabbitmq-server` file. Then, you can improve RabbitMQ itself by using HiPE.

3. Install the latest version of Erlang from `http://www.erlang.org/download.html`.

4. Install HiPE in your system.

5. Check that HiPE is correctly activated; if not, you need to install Erlang from the sources and activate it.

6. Activate Erlang HiPE in the RabbitMQ configuration file. Create the `rabbitmq.config` file with this option or add it if the file already exists:

   ```
   [
       {rabbit, [{hipe_compile, true}]}
   ].
   ```

7. Restart RabbitMQ.

8. Check that in the RabbitMQ log file there is not a warning showing that HiPE has not been activated:

   ```
   =WARNING REPORT==== 6-Oct-2013::00:38:23 ===
   Not HiPE compiling: HiPE not found in this Erlang installation.
   ```

How it works...

The watermark is the maximum memory used by RabbitMQ, by default it's 0.4 which means 40 percent of the installed physical memory. When the memory reaches the watermark the broker stops accepting new connections and messages. The watermark value is approximate; in some cases it could be overcome by the default 40 percent. Anyway, when the server has lots of RAM, you can increase the value, for example, to 60 percent, just to tolerate the spikes. With `rabbitmqctl` the change is temporary; when you modify the `rabbitmq.config` file, the option is set permanently.

The `ulimit` parameter by default is `1024`. Increase the value to increase the number of files and of sockets available to RabbitMQ.

 A too high value could impact negatively the system. Read about the `ulimit` parameter at `https://wiki.debian.org/Limits`.

Using Erlang HiPE is currently considered experimental. If it works, we can use it. In case the system is unstable, you need to disable it.

However, using it you can obtain a consistent 40 percent of CPU usage improvement of the RabbitMQ server, in case this is your bottleneck. For example, in the following screenshot, you can see the behavior of the broker in a standard configuration with a producer and a consumer on localhost:

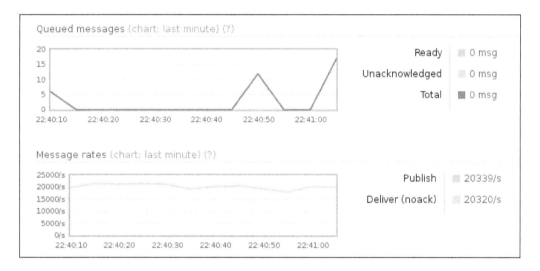

In this example, we have run both the producer and the consumer on the localhost, sending 32 byte messages for 300 seconds, letting the consumer consume all the messages in real time.

After HiPE has been activated, as shown with details in the following screenshot, the same test behaves considerably better:

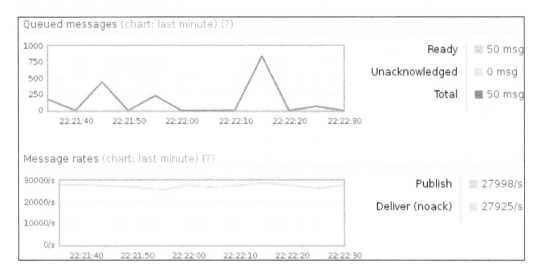

Before you activate HiPE in the RabbitMQ configuration file, you can check if your local Erlang installation has it by just invoking the `erl` command as follows:

```
# erl
Erlang R15B03 (erts-5.9.3.1) [source] [64-bit] [smp:2:2] [async-
threads:0] [hipe] [kernel-poll:false]


EshellV5.9.3.1  (abort with ^G)
1>
```

In case HiPE is present, you will see `[hipe]` among the options shown at startup.

> In Debian wheeze systems, once the latest Erlang version from the Erlang-site is downloaded, you can add the HiPE module by installing:
>
> ```
> apt-get install erlang-base-hipe
> ```
>
> The other distributions have similar packages available as well.

Otherwise, you need to install it from an external package or from the Erlang source code by downloading it from `http://www.erlang.org/download.html` and installing it; remember to specify the `--enable-hipe` option at the `configure` step.

Once RabbitMQ has been configured too (in step 6), you will notice that the restart of the server will take a long time; typically several minutes.

 In most of the Linux distributions, the default RabbitMQ configuration file is placed at `/etc/rabbitmq/rabbitmq.config`.

At this point, the RabbitMQ broker is HiPE-activated. The most demanding parts are not interpreted anymore but compiled at startup into native machine code.

You can further check that in the log file you don't see any message as follows:

```
=WARNING REPORT==== 6-Oct-2013::00:38:23 ===
Not HiPE compiling: HiPE not found in this Erlang installation.
```

 By default, on Linux RabbitMQ, log files are placed in `/var/log/rabbitmq`. You can find more information in *Chapter 12, Managing RabbitMQ Error Conditions*.

There's more...

Since HiPE is an experimental option, we discourage its usage from the beginning, given that usually the optimization effort needs to address the application side optimization and scalability.

However, by enabling it, you can reduce the CPU usage and power consumption of your servers; so this is an option you can consider when optimizing your architecture.

Improving bandwidth

Using **noAck** flag and managing the **prefetch** parameter is another client-side way to improve the performance and the bandwidth. Both noAck and prefetch are used by the consumers.

In this example, we are going to create one producer and one consumer using these parameters. You can find the source code at `Chapter08/Recipe03`.

Getting ready

You need Java 1.7 or higher and Apache maven.

How to do it...

We skip the producer code because it's the same one shown in the *Multithreading and queues* recipe. We still use the `ReliableClient` class as the base class. Let's see the consumer by performing the following steps:

1. Create a maven project and add the RabbitMQ client dependency.

2. Create a consumer main class, which reads from `args []` to manage the consumer with the following four parameters:

   ```
   threadNumber = Integer.valueOf(args[0]);
   prefetchcount = Integer.valueOf(args[1]);
   autoAck = (Integer.valueOf(args[2]) != 0);
   print_thread_consumer
   = (Integer.valueOf(args[3]) != 0);
   ```

3. Create a consumer that extends the `ReliableClient` class, and then set the `prefetch` and `noAck` parameters:

   ```
   internalChannel.basicQos(prefetch_count);
   internalChannel.basicConsume(Constants.queue, autoAck..
   ```

How it works...

The prefetch-size is ignored if the `noAck` option is set, so we divide the recipe in the following two sections:

- ▸ Prefetch
- ▸ noAck

The aim is to understand how to manage the client-side parameters to improve the performance and the bandwidth.

Prefetch

To set the prefetch, use `basicQos(prefetch_count)` (refer to step 3).

We have already seen the `channel QoS` parameter in *Chapter 1*, Working with AMQP, *Distributing Messages to Many Consumers*, where the messages are acknowledged one by one, in order to correctly load balance the messages.

The prefetch count is the maximum number of unacknowledged messages: a large value will let the client prefetch many messages in advance without waiting for the acks of the messages being processed.

As stated at the beginning of the chapter, there is not a one-for-all rule when optimizing. In fact, improving the prefetch count can be counterproductive when the per message processing time is important, and we need to distribute and balance the processing.

Well firstly, maven will compile using the following command:

`mvn clean compile assembly:single`

Then, maven will create the `rmqAckTest.jar` package.

You can now try the example, by changing the parameters to see how the message rates change. We made a test using a MacBook pro Dual Core, 4 GB RAM using the following parameters:

- ▸ For the producer, we can run the following command:

  `java -cp rmqAckTest.jar rmqexample.ProducerMain 1 100000 64000`

- ▸ For the consumer, we can run the following two tests:

  `java -cp rmqAckTest.jar rmqexample.ConsumerMain 2 50 0 0`

  `java -cp rmqAckTest.jar rmqexample.ConsumerMain 2 1 0 0`

The producer uses 1 thread for 100 seconds and 64000 bytes as the message size.

The consumer uses 2 threads, the prefetch count in `Test1` is 50 and in `Test2` is 1, `autoAck` is set to `false`, and without printing the thread number on the console.

The results for these two tests were as follows:

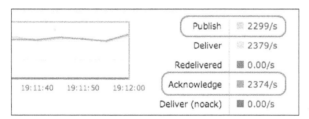

Test1

Test2

As you can see, the prefetch count can make the difference, especially when you have more consumers bound to the same queue.

NoAck

To set noAck, use `basicConsume(Constants.queue, true)`. The parameter is useful when you have a stream data or when it doesn't matter to send the acks manually.

 In the Java API, the name of the `ack` parameter of `Channel.basicConsume` as a Boolean, `autoack`, is quite misleading; setting `autoack=true` actually means that we are setting the `noAck` option.

When noAck is set, you must not call the following method:

```
internalChannel.basicAck(envelope.getDeliveryTag(), false);
```

Try to execute the test with the third parameter set to 1 (the second one is ignored) as follows:

```
java -cp rmqAckTest.jar rmqexample.ConsumerMain 1 1 1 0
```

There's more...

When optimizing messaging operations, you can obtain performance gains by acting on application-side optimizations. This is feasible both for "small" and "large" messages:

- If the size of your messages is too small, you can aggregate them manually before sending them and unpack them at the receiver side
- If the size of the messages is too large, you can try to compress the message before sending it and decompress it at the consumer side

See also

Read the article at http://www.rabbitmq.com/blog/2012/04/17/rabbitmq-performance-measurements-part-1/ and http://www.rabbitmq.com/blog/2012/04/25/rabbitmq-performance-measurements-part-2/ to understand how important every single parameter is.

Using different distribution tools

When the application needs performance, you have to choose the right distribution tool. In this example, we will show the differences between publishing a message between a mirrored queue and non-mirrored one.

Getting ready

You need Java 1.7 or higher and Apache Maven.

How to do it...

You can use the source code from the *Improving bandwidth* recipe, then you have to create a RabbitMQ cluster with two nodes.

How it works...

A cluster using HA mirrored queues is slower than a single broker. Higher the number of mirroring servers, slower the application will be because the producer can send more messages only after the message being sent is stored to all the mirrors.

 That's not as bad as it might seem. On one side, the distribution toward the nodes of the cluster is performed in parallel, so the overhead does not grow linearly with the number of nodes. On the other side, the replication is usually limited to two or three replicas at the most, as we saw in *Chapter 7, Developing High-availability Applications*.

We performed a test using the following environment:

- ▶ `https://www.digitalocean.com/` as cloud
- ▶ Two Debian machines in a RabbitMQ cluster, as shown in the following screenshot:

- ▶ One Debian machine with the same characteristics as the Java client

The tests performed are as follows:

▶ **Test 1**: Create a mirror (as we saw in *Chapter 7, Developing High-availability Applications*) using the configuration, as shown in the following screenshot:

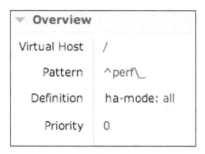

So. the cluster will mirror all queues with the `perf_` prefix.

The producer is run with the following command:

```
java -cp rmqAckTest.jar rmqexample.ProducerMain 1 100000 640
```

The consumer is run with the following command:

```
java -cp rmqAckTest.jar rmqexample.ConsumerMain 1 0 0 0
```

The clients exchange messages through the `perf_queue_08/03` queue, on which the observed performance is as follows:

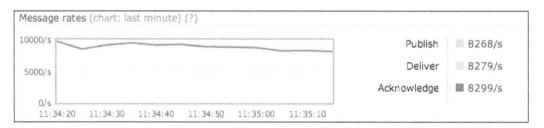

▶ **Test 2**: We removed the HA policy and tried again. The result, in this case, was similar to the following screenshot:

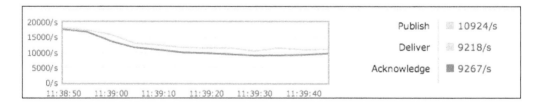

▶ **Conclusion**: By using small-sized messages, we have amplified the differences. For larger messages, they are much less evident. In Test 2, we have observed about 2.000 mgs/s more than Test 1, but as you can see, the rate dropped because the producer is faster than the consumer.

 As a general rule, high availability has a negative impact on performance. So, whenever it's not mandatory, it's better to leave it off.

In this example, we have gone to try the highest performance and in this context, we have seen the impact of queue mirroring. If we need a level of replication, but without the strict requirements of mirroring, it is possible to use a shovel plugin or simply publish the message to two independent brokers in parallel. In this example, the messages aren't persistent and don't use the **tx-transaction**.

 tx-transactions kill the performance, especially when you try to commit each single message, because it has to wait the disk flush time for each message.

There's more...

Finding the right compromise between performance and reliability is very hard because there are lots of variables inside a distributed application. A typical error is trying to optimize each single application flow losing scalability or eventually high-availability benefits. This chapter presents extreme situations, but as we have seen, there are margins for improvement.

See also

Performance is a hot topic in the RabbitMQ mailing list. You can search and find a lot of useful information in the archives at `http://rabbitmq.markmail.org/`.

9
Extending RabbitMQ Functionality

In this chapter we will present some of the available plugins of RabbitMQ. Then, we will show how to develop new plugins using real-world examples.

- ▸ Enabling and configuring the STOMP plugin
- ▸ Managing a RabbitMQ cluster
- ▸ Monitoring Shovel status
- ▸ Developing new plugins – attaching to a relational database with ODBC

Introduction

RabbitMQ is an extensible platform, thanks to the plugin infrastructure. It provides many general-purpose plugins, part of which were already explained in the previous chapters. For example, Federation and Shovel plugins have been explained in detail in *Chapter 7, Developing High-availability Applications*.

In the following recipes, we will show how to use other available plugins and how to develop new plugins to create custom extensions for RabbitMQ.

Enabling and configuring the STOMP plugin

We have already seen how to use plugins from the very beginning of this book. The management plugin itself is a plugin actually. However, our intent is to show further uses of plugins.

In this recipe, we will see how to enable the STOMP plugin and the further possibilities it provides to RabbitMQ.

With **Simple** (or **Streaming**) **Text Orientated Messaging Protocol** (**STOMP**), (`http://stomp.github.io/`), RabbitMQ increases its language interoperability. With the plugin installed, the RabbitMQ broker can operate not only with the AMQP protocol, but with STOMP as well.

Getting ready

For this recipe you just need the latest version of RabbitMQ.

How to do it...

In order to run this recipe, you need to perform the following steps:

1. From root (Linux) or from the **RabbitMQ Command Prompt** (Windows), check the current plugin state with the following command:

   `rabbitmq-plugins list`

2. Then enable the STOMP plugin by issuing the following command:

   `rabbitmq-plugins enable rabbitmq_stomp`

3. Restart RabbitMQ with the following command:

   `service rabbitmq-server restart`

At this point, you can try to submit some STOMP messages using Netcat, the `nc` command (`http://en.wikipedia.org/wiki/Netcat`), from the terminal session (Windows users can use `telnet`):

4. From a shell prompt, type the following `nc` command:

   ```
   nc localhost 61613
   CONNECT

   ^@
   SEND
   Destination:/queue/test
   This the 1st stomp message

   ^@
   ```

 Note that ^@ stands for the character obtained with the combination *CTRL* + @, equivalent to the character with an ASCII code equal to zero.

5. From a second terminal, type the following `nc` command:

```
nc localhost 61613
CONNECT

^@
SUBSCRIBE
destination:/queue/test

^@
```

How it works...

By listing the available plugins (step 1), we get an output as shown in the following screenshot:

```
                              root@localhost:~                              ×
 File   Edit   View   Search   Terminal   Help
[root@localhost ~]# rabbitmq-plugins list
[e] amqp_client                             3.1.5
[ ] cowboy                                  0.5.0-rmq3.1.5-git4b93c2d
[ ] eldap                                   3.1.5-gite309de4
[e] mochiweb                                2.7.0-rmq3.1.5-git680dba8
[ ] rabbitmq_amqp1_0                        3.1.5
[ ] rabbitmq_auth_backend_ldap              3.1.5
[ ] rabbitmq_auth_mechanism_ssl             3.1.5
[ ] rabbitmq_consistent_hash_exchange       3.1.5
[ ] rabbitmq_federation                     3.1.5
[ ] rabbitmq_federation_management          3.1.5
[ ] rabbitmq_jsonrpc                        3.1.5
[ ] rabbitmq_jsonrpc_channel                3.1.5
[ ] rabbitmq_jsonrpc_channel_examples       3.1.5
[E] rabbitmq_management                     3.1.5
[e] rabbitmq_management_agent               3.1.5
[ ] rabbitmq_management_visualiser          3.1.5
[ ] rabbitmq_mqtt                           3.1.5
[E] rabbitmq_shovel                         3.1.5
[ ] rabbitmq_shovel_management              3.1.5
[ ] rabbitmq_stomp                          3.1.5
[ ] rabbitmq_tracing                        3.1.5
[e] rabbitmq_web_dispatch                   3.1.5
[ ] rabbitmq_web_stomp                      3.1.5
[ ] rabbitmq_web_stomp_examples             3.1.5
[ ] rfc4627_jsonrpc                         3.1.5-git5e67120
[ ] sockjs                                  0.3.4-rmq3.1.5-git3132eb9
[e] webmachine                              1.10.3-rmq3.1.5-gite9359c7
[root@localhost ~]# []
```

These are all the built-in plugins available. The plugins marked with empty brackets are not installed. The ones marked with `[E]` are plugins that are explicitly installed. The ones marked with `[e]` are implicitly installed, that is, plugins are automatically installed because of their dependencies on other plugins.

After we install the STOMP plugin, we need to restart the broker to let the change be effective. Once the plugin is activated, you can use the simple, textual STOMP protocol to send a message to the queue `test` (step 4). This will give what you can see in the following screenshot:

```
                                terminal 1                              _ □ ✕

 File   Edit   View   Search   Terminal   Help
 [root@localhost ~]# nc localhost 61613
 CONNECT

 ^@
 CONNECTED
 session:session-FaKDfHq5DwZwnl5YBdlitA
 heart-beat:0,0
 server:RabbitMQ/3.1.5
 version:1.0

 SEND
 destination:/queue/test

 this is the 1st stomp message

 ^@

 ▯
```

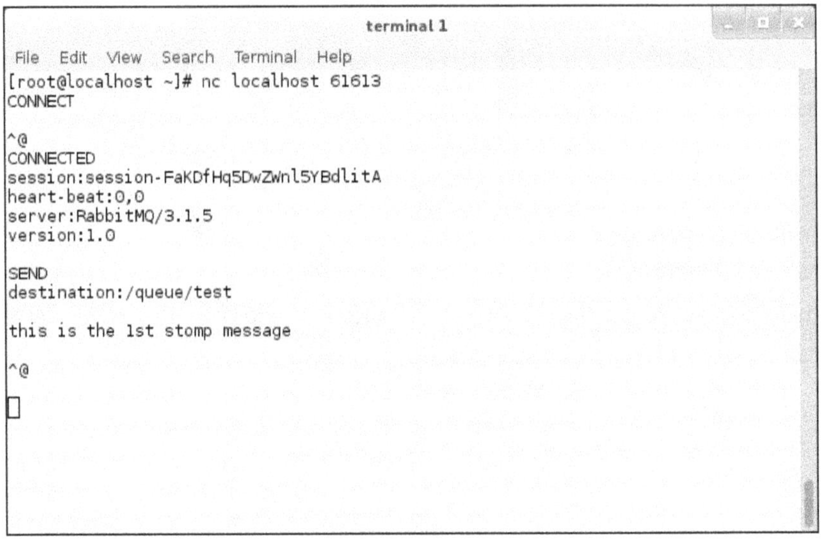

On another terminal, we can start consuming messages from the queue `test` (step 5). This will give what you can see in the following screenshot:

```
                                terminal 2                              _ □ ✕

 File   Edit   View   Search   Terminal   Help
 [root@localhost ~]# nc localhost 61613
 CONNECT

 ^@
 CONNECTED
 session:session-1Cs8OltsqKKTHriK1WanLg
 heart-beat:0,0
 server:RabbitMQ/3.1.5
 version:1.0

 SUBSCRIBE
 destination:/queue/test

 ^@
 MESSAGE
 destination:/queue/test
 message-id:Q_/queue/test@@session-1Cs8OltsqKKTHriK1WanLg@@1
 content-length:31

 this is the 1st stomp message

 ▯
```

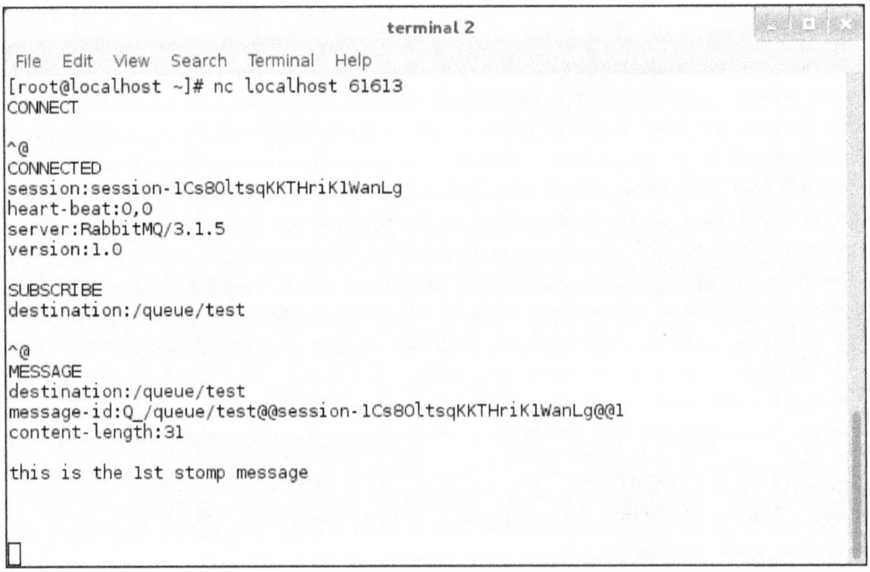

Note that the text from MESSAGE to the end has not been typed—that's the actual received message.

There's more...

In this recipe, we have used the standard STOMP configuration. However, it is possible to customize the STOMP port, SSL usage, and STOMP users by adding the proper configuration options in the RabbitMQ configuration file.

You can find more details at `http://www.rabbitmq.com/stomp.html`.

 Note that STOMP is not exactly the same as Web-Stomp, which is STOMP encapsulated within WebSockets. See the *Developing web monitoring applications with STOMP* recipe in *Chapter 5, Using RabbitMQ in Web Applications*. The two protocols are not interoperable.

Furthermore, it's possible to send other kinds of messages with STOMP. It's possible to send RPC messages with temporary queues, and send/receive messages to/from exchanges.

See also

In this recipe, we have shown how to interact with RabbitMQ and its STOMP plugin by using Netcat as a textual client. However, this is not the typical way to use STOMP at the client side. There are a lot of STOMP client APIs available.

You can find a list of the available client libraries at `http://stomp.github.io/implementations.html`.

Managing a RabbitMQ cluster

When deploying a large RabbitMQ cluster, the management plugin will impose a sensitive overhead on the cluster operations. In this recipe, we are showing how to relieve this overhead.

Getting ready

In order to experiment this recipe, you need a RabbitMQ cluster with at least two nodes. If they have the management plugin already installed, you need to remove it.

How to do it...

1. Install the management plugin on one of the nodes using the following command:

   ```
   rabbitmq-plugins enable rabbitmq_management
   ```

2. Install the management agent plugin on all the other nodes using the following command:

   ```
   rabbitmq-plugins enable rabbitmq_management_agent
   ```

How it works...

After performing these steps, you will be able to monitor the entire cluster from the first node only. The other nodes will have the agents updating the status of the console of the first node, but you won't be able to access port 15672 on them.

Typically, you will install the full management plugin on a couple of nodes, such as frontend or management nodes, and the management agent plugin on the rest of them.

Monitoring Shovel status

We have already seen in *Chapter 7, Developing High-availability Applications*, how to use the Shovel plugin. In this recipe, we are going to see how to monitor its correct behavior using an appropriate plugin. This is an extension of the `rabbitmq_management` plugin.

Getting ready

To test this recipe, we need two RabbitMQ brokers running. We will refer to them in this recipe as `rabbit@node01` and `rabbit@node02`.

How to do it...

We assume that the two brokers are already running on their respective nodes. We will start configuring the broker on `node01` with the configuration file that you can eventually copy and adapt from the one available in the book repository in `Chapter07/Recipe05`:

1. In the RabbitMQ configuration file, generally `/etc/rabbit/rabbitmq.config`, insert the Shovel configuration:

   ```
   [{rabbitmq_shovel,
   [ {shovels, [ {my_books_shovel,
     [
       {sources, [ {broker, "amqp://node02"}]}
     , {destinations, [ {broker, "amqp://"}]}
   ```

```
     , {queue, <<"myBooksQueueCopy">>}
     , {prefetch_count, 10}
     , {reconnect_delay, 5}
       ]}]}].
```

2. Activate the needed plugins by typing the commands:

    ```
    rabbitmq-plugins enable rabbitmq_management
    rabbitmq-plugins enable rabbitmq_shovel
    rabbitmq-plugins enable rabbitmq_shovel_management
    ```

3. Restart the broker for the changes to take effect using the following command:

    ```
    service rabbitmq-server restart
    ```

4. Access the RabbitMQ management interface at the following URL:

    ```
    http://node01:15672/
    ```

5. From the management interface, create the `myBooksQueueCopy` queue on both `node01` and `node02`.

6. In the Management interface, navigate to **Admin | Shovel Status**.

How it works...

When we activate the Shovel plugin, it is possible to monitor its correct behavior by using the `rabbitmq_shovel_management` plugin. Once we activate the plugin, since the redirected queue does not exist, you will be able to see an error condition. This is shown in the following screenshot:

As soon as we create the queue (step 5), the page is automatically updated within 5 seconds:

On this page, RabbitMQ will allow us to monitor all the configured Shovels.

Developing new plugins – attaching to a relational database with ODBC

In the previous recipes, we have seen how to use some existing plugins and what can be performed with them. Now, we are going to see how to develop a new plugin for our custom usage.

Note that this should not be a typical practice. Most of the operations can, and usually are, performed with the RabbitMQ/AMQP client API.

However, it is sometimes necessary to perform some customization in the broker heart. This is done for optimization or because of the need to strictly monitor the broker actions themselves.

In this recipe, we will show how to let RabbitMQ consume messages and store them in a relational database by using the ODBC driver.

Getting ready

For this recipe, we need RabbitMQ and PostgreSQL. However, this recipe can be easily adapted to almost every relational database, given the widespread availability of ODBC drivers. Furthermore, we need to have Mercurial (http://mercurial.selenic.com/) installed, which is the versioning system used by RabbitMQ, to check out the RabbitMQ sources.

Here, the recipe is using Linux. But with limited modifications, it should work on Windows as well.

How to do it...

We will start installing and configuring PostgreSQL now.

1. Install PostgreSQL and its ODBC driver. For example, on Linux distributions (such as RedHat, CentOS, Fedora and others) that are based on yum, issue these commands, as root.

   ```
   yum install postgresql-server
   yum install postgresql-odbc
   postgresql-setup initdb
   service postgresql start
   ```

2. Create a new user (password rmq_plugin_password) and a new database for usage from the RabbitMQ plugin using the following command:

   ```
   su postgres
   createuser --no-superuser --no-createdb --no-createrole --
     pwpromptrmq_plugin_user
   createdb --owner rmq_plugin_userrmqdb
   ```

3. Allow the newly created user to access the given database (from localhost only) by adding the following lines to the end of the file /var/lib/pgsql/data/pg_hba.conf:

   ```
   # TYPE  DATABASE  USER                    ADDRESS        METHOD
   local   rmqdb     rmq_plugin_user                        md5
   host    rmqdb     rmq_plugin_user 127.0.0.1/32  md5
   ```

4. Reload the PostgreSQL configuration by issuing as root:

   ```
   service postgresql reload
   ```

5. At this point, you should be able to connect to PostgreSQL locally, for example, by issuing the following command:

   ```
   psql --username=rmq_plugin_user --dbname=rmqdb
   ```

6. Complete the setup of the configuration of the ODBC driver to access this database. To do this, insert the following lines into the file `/etc/odbc.ini`:

```
[rmqDSN]
Driver = PostgreSQL
Description = PostgreSQL data source for RabbitMQ
Servername = localhost
Port = 5432
Protocol = 8.4
Database = rmqdb
```

7. We can now test the ODBC connection with the `isql` command, but we can test it directly with Erlang too. This can be done by invoking `erl` and issuing the following commands:

```
odbc:start()
odbc:connect("DSN=rmqDSN;UID=rmq_plugin_user;PWD=rmq_plugin
    _password", [])
```

At this point, we have prepared what's needed to see the plugin that we are developing in action. Now, we will see how to enable the metronome plugin. This is presented in the official RabbitMQ documentation (`http://www.rabbitmq.com/plugin-development.html`:

8. Checkout the RabbitMQ development source's tree using the following command:

```
cd $HOME
hg clone http://hg.rabbitmq.com/rabbitmq-public-umbrella
cd rabbitmq-public-umbrella
make co
```

9. Build the metronome plugin using the following command:

```
cd rabbitmq-metronome
make
```

10. Install the plugin with its requisites in the development broker.

```
cd ../rabbitmq-server
mkdir plugins
cd plugins
ln -s ../../rabbitmq-metronome
ln-s../../rabbitmq-erlang-client
```

11. In order to avoid the override of the eventual production installation, even though we are not using the `root` user, we need to stop the eventual production server, set some environment variables, and create the corresponding directories to let RabbitMQ start as standard user:

```
export RABBITMQ_LOG_BASE=$HOME/rmq/log
export RABBITMQ_MNESIA_BASE=$HOME/rmq/mnesia
```

```
export RABBITMQ_ENABLED_PLUGINS_FILE=$HOME/rmq/enabled_plugins
mkdir -p $RABBITMQ_LOG_BASE $RABBITMQ_MNESIA_BASE
```

12. We can now enable the development plugin.

```
cd $HOME/rabbitmq-public-umbrella/rabbitmq-server
scripts/rabbitmq-plugins list
scripts/rabbitmq-plugins enable rabbitmq_metronome
```

13. And finally, start the development server interactively:

```
scripts/rabbitmq-server
```

14. At this point, we are ready to show how to develop a new plugin, which enables interfacing RabbitMQ with a third-party database by using ODBC. You can find the full source of the recipe plugin in the book source archive in the path `Chapter09/Recipe04`. You can copy the recipe source directory from `Chapter09/Recipe04/rabbitmq-odbctap` into the RabbitMQ development tree, `rabbitmq-public-umbrella`, which we checked out at step 8. In this way, you will obtain a source tree similar to the following screenshot:

15. To start developing a new plugin, copy and rename these files from an existing plugin as `rabbitmq-metronome` or `rabbitmq-shovel`. The makefile will be left unmodified.

16. Adapt the `package.mk` file to reflect the needed dependencies and the test modules.

17. Edit `rabbitmq_odbctap.app.src`. This file contains the resources used by the Erlang project, both general and custom that in our example must be set to contain the following configuration:

```
{application, rabbitmq_odbctap,
  [{description, "Embedded Rabbit ODBC tap"},
   {vsn, "0.0.0"},
   {modules, []},
   {registered, []},
   {mod, {rabbit_odbctap, []}},
   {env, [{dsn,"DSN"},
          {user,"guest"},
          {password, "guest"},
          {queue, "tapped_queue"}
         ]},
   {applications, [kernel, stdlib, rabbit, amqp_client]}]}.
```

18. Customize `rabbit_odbctap.erl`, the entry point for the module, to let it start, configure the plugin itself, and stop.

19. Customize `rabbit_odbctap_sup.erl`, the source for the Erlang supervisor node, by defining the callbacks for the Erlang supervisor behavior.

20. Implement the plugin logics in `rabbit_odbctap_worker.erl`. This is the entry point of the actual module, which in our specific case will connect to the RabbitMQ host broker internally, bind to a queue, and consume the contained messages to a specific database table via ODBC.

21. Data definitions (that is, Erlang records) common to many modules can be placed in the include directory. For example, in `rabbit_odbctap.hrl`, you can find the definition of the Erlang record `odbctap_config`, which is used by both `rabbit_odbctap.erl` and `rabbit_odbctap_worker.erl`.

22. Prepare one or more test modules. In our example, you can find just a skeleton in `rabbit_odbctap_tests.erl`.

23. Compile the plugin and perform the automated tests.

```
cd $HOME/rabbitmq-public-umbrella/rabbimq-odbctap
make
make test
```

24. Install the plugin in the development server.

```
cd ../rabbitmq-server/plugins
ln -s ../../rabbitmq-odbctap .
```

25. After setting the environment as shown in step 11, we can enable the plugin and (re) start the server.

```
cd $HOME/rabbitmq-public-umbrella/rabbitmq-server
scripts/rabbitmq-plugins enable rabbitmq_odbctap
scripts/rabbitmq-server
```

How it works...

In order to put this recipe in action, we have started configuring a PostgreSQL database, and its corresponding ODBC driver setup appropriately. This is shown in the first part of the recipe.

Then, we have shown how to enable the example template plugin `rabbitmq_metronome`, which is provided in the development tree of RabbitMQ itself. If everything goes fine, the development broker started at step 13 should be up and running as our standard user; and in the log file `$HOME/rmq/log/rabbit@localhost.log`, you will be able to get the following as final lines:

```
=INFO REPORT==== 21-Oct-2013::03:28:50 ===
Server startup complete; 2 plugins started.
 * amqp_client
 * rabbitmq_metronome
```

This is the starting point for the development of a new plugin that uses the Erlang ODBC client to monitor some given queues in the configured PostgreSQL database. This is shown in the third part of this recipe.

To implement our plugin, we have chosen to follow a common approach, that is, to use modules implementing Erlang behaviors. In particular, Erlang encourages using an architecture where the application (our plugin) is decoupled by the launcher (RabbitMQ itself) by using a supervisor. For more details on these aspects, see `http://www.erlang.org/doc/man/supervisor.html` and `http://www.erlang.org/doc/design_principles/sup_princ.html`.

The only customization needed for the supervisor is that it must be capable of getting the configuration from the startup module and passing it to the worker module. For this purpose, the startup module (`rabbit_odbctap.erl`) contains the code that reads the configuration itself.

```
read_config() ->
    {ok, Dsn}      = application:get_env(dsn),
    {ok, User}     = application:get_env(user),
    {ok, Password} = application:get_env(password),
    {ok, Queue}    = application:get_env(queue),
    Config = #odbctap_config{
            dsn = Dsn,
            user = User,
```

```
              password = Password,
              queue = Queue},
    Config.
```

The calls to `application:get_env/1` let the module automatically access the definitions that can be set as follows:

- In the Erlang resource file, `rabbitmq_odbctap.app.src`, the env key is set at compile time:

  ```
  . . .
    {env, [{dsn,"DSN"},
           {user,"guest"},
           {password, "guest"},
           {queue, "tapped_queue"}
          ]},
    ...
  ```

- In the `rabbitmq.config` file, read at runtime, which follows the typical RabbitMQ (that is, Erlang) format , specifying a `rabbitmq_odbctap` key as in the following code:

  ```
  [{rabbitmq_odbctap,
     [{dsn, "rmqDSN"},
      {user,"rmq_plugin_user"},
      {password, "rmq_plugin_password"},
      {queue, "tapped_queue"}]}].
  ```

You can find the complete files in both the forms in the book archive directory.

The `rabbit_odbctap_worker.erl` worker module implements the `gen_server` behavior. You can find more information at `http://www.erlang.org/doc/design_principles/gen_server_concepts.html`.

In our module, once the module is started, `init/2` gets called. Then it connects to the defined ODBC connector by using the ODBC client Erlang API `http://www.erlang.org/doc/apps/odbc/` and fires up the RabbitMQ consumer. It's important to note that in this case the embedded RabbitMQ client connects to the local broker, embedded within the same Erlang virtual machine, with the call:

```
{ok, Connection} = amqp_connection:start(#amqp_params_direct{})
```

In this way, the plugin uses the internal Erlang connections and protocols with a much more efficient messaging solution, since the marshalling from/to the AMQP wire protocol is totally avoided.

To log some information to the RabbitMQ log file, you can use `rabbit_log:info/2` and similar calls, as shown in the `init/2` definition.

By using the `gen_server` behavior, the messages to the broker are not consumed from a receive block (http://www.rabbitmq.com/erlang-client-user-guide.html, see the *Subscribing To Queues* section), but by using the `handle_info/2` callback that we are reporting here:

```
handle_info({#'basic.deliver'{delivery_tag = Tag},
    #amqp_msg{payload = Payload}},State = #state{channel=Channel,
        odbcHandle = Ohandle}) ->
Query = "INSERT INTO rmqmessages VALUES ('now','"++
    binary_to_list(Payload)  ++"')",
{updated, _} = odbc:sql_query(Ohandle, Query),
amqp_channel:cast(Channel, #'basic.ack'{delivery_tag = Tag}),
{noreply, State};
```

This is the heart of the recipe. Each consumed message is stored in the database opened at initialization time within RabbitMQ itself.

There's more...

It's important to remember that developing a plugin is the last choice. It's much better to develop an external application by using an AMQP client library.

 A plugin bug can compromise the RabbitMQ stability.

So, only when extreme integration and possibly performance is needed, you can evaluate whether to develop a plugin.

10

RabbitMQ on AWS

In this chapter we will cover:

- ▶ Using RabbitMQ EC2 instances
- ▶ Creating a master image
- ▶ Creating a cluster with two EC2 instances
- ▶ Using AWS Load Balancing in front of a RabbitMQ cluster
- ▶ Configuring EC2 dynamic bind
- ▶ Dealing with load spikes and resource optimization in the cloud

Introduction

Amazon Web Services (**AWS**), available at `http://aws.amazon.com/`, is one of the most used cloud service providers. In this chapter, we will see how to use RabbitMQ on AWS, starting with a simple RabbitMQ configuration and finishing with an auto-scaling RabbitMQ cluster. You need to create an account on AWS to execute the examples.

Amazon free usage tier is a good opportunity to try AWS (`http://aws.amazon.com/free/`).

In the free tier, you can run a micro EC2 instance and some other services for free for a limited amount of time. However, you will be paying for storage and other services. You can configure a billing alert to keep under control your budget. For more information go to `http://docs.aws.amazon.com/AmazonCloudWatch/latest/DeveloperGuide/monitor_estimated_charges_with_cloudwatch.html`.

We are going to use the following AWS products:

- **Amazon Elastic Compute Cloud** (**EC2**): This is the AWS service that provides the computing capacity as instances (`http://aws.amazon.com/ec2/`)

- **Elastic Load Balancing**: This allows us to uniformly distribute the workload among the running instances (`http://aws.amazon.com/elasticloadbalancing`)

- **Amazon Virtual Private Cloud** (**VPC**): This allows us to create a virtual private network among the running instances (`http://aws.amazon.com/vpc`)

- **Amazon CloudWatch**: This allows us to monitor and trigger automatic scaling actions on the running instances (`http://aws.amazon.com/cloudwatch`)

- **Amazon Auto Scaling**: This allows us to add or remove new instances automatically (`http://aws.amazon.com/autoscaling`)

As the operating system to be installed on the EC2 instances, we have chosen Ubuntu 12.04.3 LTS Cloud, which you can find among the **Amazon Machine Image** (**AMI**) available at `http://cloud-images.ubuntu.com/locator/ec2`.

Using RabbitMQ EC2 instances

In this recipe we will show how to create an Amazon EC2 instance, and how to configure RabbitMQ to be accessible from the Internet.

Getting ready

You need an Amazon AWS account.

How to do it...

You should have basic knowledge of Amazon EC2; however, we will explain the steps in detail:

1. Log in to AWS and go to the EC2 console.
2. Choose a region, for example, Northern Virginia, from the available regions:

3. Create a key pair to use in these examples, by navigating to **Network & Security |
Key Pairs | Create Key Pair** as shown in the following screenshot:

The browser will automatically download the `rabbitmqkey.pem` file, which contains
the private key.

4. You need to change the permission of the `rabbitmqkey.pem` file by executing the
following command:

```
chmod 400 rabbitmqkey.pem
```

5. Create your security group by navigating to **Network & Security | Security Groups |
Create Security Group**:

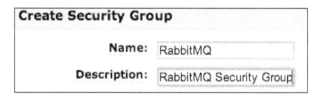

6. Add the following port to the RabbitMQ security group:

   ❑ 22: SSH port

   ❑ 5672: RabbitMQ port

   ❑ 15672: Web management port

7. Go to `http://uec-images.ubuntu.com/releases/12.04.3/release/` and
choose your AMI from the region used until now, and click on the **Launch** button:

8. Follow the wizard, taking care to use the same region, otherwise you will find neither your key pair nor your security group (steps 3 and 4).

9. Connect via SSH to the instance using the following command:

   ```
   ssh -i rabbitmqkey.pemubuntu@[public DNS instance]
   ```

   If you are using Windows, please follow the guide at `http://docs.aws.amazon.com/AWSEC2/latest/UserGuide/putty.html`.

Once connected, you can install RabbitMQ as always.

10. Add the RabbitMQ Debian repository to the apt-get configuration file by appending to the `/etc/apt/sources.list` file the following code line:

    ```
    deb http://www.rabbitmq.com/debian/ testing main
    ```

11. Install RabbitMQ with the following commands:

    ```
    wget http://www.rabbitmq.com/rabbitmq-signing-key-public.asc
    sudo apt-key add rabbitmq-signing-key-public.asc
    sudo apt-get -qy update
    sudo apt-get -qy install rabbitmq-server
    ```

12. Stop RabbitMQ; now fix the hostname issue by executing the following command:

    ```
    sudo -s
    echo"rabbit1"> /etc/hostname
    echo"127.0.0.1 rabbit1">> /etc/hosts
    hostname -F /etc/hostname
    ```

13. Install the RabbitMQ web management console.

14. Restart RabbitMQ.

How it works...

The AWS infrastructure contains different regions, and when you use AWS EC2 you must choose one, where your instances will be run. In this recipe we use the Northern Virginia region (step 2), and then create a key pair that you will use for SSH connections.

 It is possible to import an existing key if needed. See `http://docs.aws.amazon.com/AWSEC2/latest/UserGuide/ec2-key-pairs.html#how-to-generate-your-own-key-and-import-it-to-aws`.

You need to create a security group to open the TCP ports accessible from the Internet; this is a generic firewall configuration that we will use for our instance. We open the SSH port (22), the RabbitMQ port (5672), and the RabbitMQ web management port (15672).

Well, the AWS environment is ready and we can install it on our machine: at `http://uec-images.ubuntu.com/releases/12.04.3/release/` you can find some Ubuntu cloud AMIs.

 Choose the Northern Virginia AMI region to use the key pairs and security group previously created (steps 3 and 4).

Once you click on the **Launch** button (step 6) and execute the AWS instance creation wizard, you just need to select the security group configured in step 5.

The Ubuntu instance will start when the wizard finishes. Check the instance status on the AWS EC2 console:

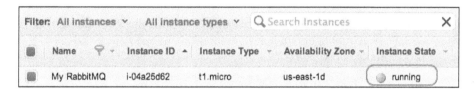

In order to use SSH connections, you need the public hostname; you can take it on the instance details, specified as **public DNS**. Alternatively, you can directly click on the **Connect** button as shown in the following screenshot:

We can use the SSH connection when the instance is ready by executing the following command:

```
ssh -i rabbitmqkey.pem ubuntu@[public DNS instance]
```

 The default username is `ubuntu` for the images created from the AMI used in the recipe. Usually the default username (that can be `root`, `admin`, `ec2-user` and more) is specified in the AMI description.

The result is shown in the following screenshot:

```
amazon $ pwd
/Users/gabriele/Documents/amazon
amazon $ chmod 400 rabbitmqkey.pem
amazon $ ssh  -i rabbitmqkey.pem ubuntu@ec2-54-242-253-34.compute-1.amazonaws.com
Welcome to Ubuntu 12.04.3 LTS (GNU/Linux 3.2.0-54-virtual x86_64)
```

Now you can follow steps 10 and 11 to install RabbitMQ, and then we need to fix the hostname issue as per step 11. In fact the EC2 hostname can change at each restart.

 Remember that RabbitMQ uses a short hostname to name the broker.

At this point you can access the RabbitMQ instance in the cloud using the public hostname provided by AWS. For web management it is something like `http://ec2-54-242-253-34.compute-1.amazonaws.com:15672/`.

Clearly this address will change from time to time.

There's more...

Using the instance public IP isn't a good practice, because the public IP can change if you restart the instance, and you need to reconfigure all your clients. You can use the Amazon Elastic IP by navigating to **EC2 console** | **Elastic IPs** | **Allocate New Address**, as shown in the following screenshot:

Then associate your static IP to the instance by selecting **Associate Address**. After a few seconds the static IP replaces the instance IP.

 It's possible to use the dynamic DNS services if you want to avoid Elastic IPs.

Read the AWS pricing policies for each of the available products at `http://aws.amazon.com/products`. For further convenience in the estimation of the cost of AWS resources, it's possible to use the AWS Simple Monthly Calculator too, at `http://calculator.s3.amazonaws.com/calc5.html`.

Creating a master image

Once your RabbitMQ instance is customized, you can create a master image, that is, a personal AMI (`http://docs.aws.amazon.com/AWSEC2/latest/UserGuide/AMIs.html`), so you won't need to reinstall your software from scratch, in case you need more instances similar to the ones already running.

Getting ready

You need a RabbitMQ instance by following the *Using RabbitMQ EC2 instances* recipe.

How to do it...

This recipe is quite easy, but it's important for the next recipes. Perform the following steps for this recipe:

1. Open the EC2 console.
2. Select your RabbitMQ instance.
3. From the **Action** menu, select the **Create Image** entry. It opens a form as shown in the following screenshot:

4. Go to **EC2 Console | Images | AMIs** to check your image. After it's completed, you can launch other instances cloned from the master.

How it works...

Creating a master image will help us to create a RabbitMQ cluster. In this way all the cloned instances will share the same `.erlang.cookie` file, the same Erlang version, and the same RabbitMQ version.

Furthermore, by freezing the versions of the software stack within the image, it's easy to add more instances coherent with the existing cluster.

See also

For more information about images read the documentation at `http://docs.aws.amazon.com/AWSEC2/latest/UserGuide/creating-an-ami.html`.

Creating a cluster with two EC2 instances

In this recipe we are going to create a RabbitMQ cluster on Amazon AWS. We will create the cluster inside a VPC (`http://aws.amazon.com/vpc/`). A VPC is a private network on the cloud, protected from the outside network, so we don't need to configure any firewall.

The schema we are going to create contains one VPC with the following two subnets:

- A public subnet accessible from the Internet
- A private subnet that contains two machines configured as a RabbitMQ cluster

  - The subnets are shown in the following diagram:

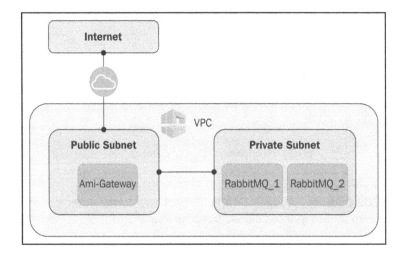

Getting ready

You need an AWS account.

How to do it...

In order to create the schema, AWS provides a ready VPC template that you can use. Perform the following steps:

1. Open the AWS console and go to the VPC dashboard.

2. Create your VPC, follow the wizard, and select the second option, which is shown in the following screenshot:

> ⦿ **VPC with Public and Private Subnets**
>
> In addition to containing a public subnet, this configuration adds a private subnet whose instances are not addressable from the Internet. Instances in the private subnet can establish outbound connections to the Internet via the public subnet using Network Address Translation.

3. Open the SSH port on the private subnet Security Group to allow the connection from the EC2 NAT instance (that is, the AMI gateway in the previous screenshot):

ALL	
Port (Service)	**Source**
ALL	sg-7b0db514
TCP	
Port (Service)	**Source**
22 (SSH)	0.0.0.0/0

4. Copy your key pair to the EC2 NAT instance using the following command:

    ```
    scp -i rabbitmqkey.pem rabbitmqkey.pem ec2-
      user@yourstaticip:/tmp
    ```

    Store the key locally.

5. Launch two RabbitMQ instances from your image repository as we saw in the *Creating a master image* recipe

6. Bind the RabbitMQ instances to the private subnet.

7. Now you can configure the RabbitMQ cluster as we saw in the *Creating a simple cluster* recipe in *Chapter 6, Developing Scalable Applications*.

How it works...

To create a RabbitMQ cluster on AWS, you can use two public EC2 instances, but we prefer to use a VPC to protect the instances. The VPC ensures the right security and contains some important features to easily create a cluster:

- **Fixed IPs**: These IPs are different from the public EC2 instances that take random IPs; the machines inside a VPC always maintain the same internal IP.

- **No need to configure the internal firewall**: By default, the security group inside the VPC allows the connection between the machines without restrictions that is optimal for the cluster.

- **Improved security**: It's not possible to access the VPC instances from the Internet directly. You need a VPN or a public gateway to access them.

Well, on the last wizard (step 2) you will have two subnets, one EC2 NAT instance and one Elastic IP as shown in the following screenshot:

Two Subnets

      Public Subnet: 10.0.0.0/24 (251 available IPs)
    Availability Zone: No Preference
      Private Subnet: 10.0.1.0/24 (251 available IPs)
    Availability Zone: No Preference

Additional subnets can be added after the VPC has been created.

One NAT Instance with an Elastic IP Address

      Instance Type: m1.small
      Key Pair Name: rabbitmqkey

Note: Instance rates apply. View rates.

The wizard will automatically select the availability zone for the subnets. Visit `http://docs.aws.amazon.com/AWSEC2/latest/UserGuide/using-regions-availability-zones.html` for more details.

After finishing the wizard, you will be able to connect to the EC2 NAT instance via SSH using the public IP we just created (step 4).

Now you can launch two RabbitMQ instances from your image (step 5), and choose the VPC private network in this wizard step:

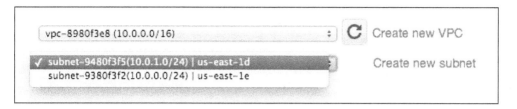

After completing the step 6 you should have a situation as shown in the following screenshot:

In our case the internal machines have the following IPs:

▸ 10.0.1.132

▸ 10.0.1.174

You can access the two private EC2 instances only from the EC2 NAT instance machine, using:

```
[ec2-user@natmachine key]$ sudo ssh -i rabbitmqkey.pem
  ubuntu@10.0.1.132   (or 10.0.1.174).
```

Inside the VPC, it's possible to use VPC's private DNS and create the cluster you have to execute by using the following code:

```
rabbitmqctl join_cluster rabbit@ip-10-0-1-132
Clustering node 'rabbit@ip-10-0-1-174' with 'rabbit@ip-10-0-1-132'
  ...
...done.
```

Ok, now your private cluster is ready, and you can use it only from the public subnet.

As you know the RabbitMQ cluster must have the same `.erlang.cookie` file. If you use a master EC2 image, you don't need to change anything because the file is stored on the image. If you don't use the image, remember to copy the `.erlang.cookie` file (as we saw in the *Creating a simple cluster* recipe in *Chapter 6, Developing Scalable Applications*).

There's more...

In order to access the VPC from the outside, it's needed either to have a proxy on the gateway host, or to use a VPN so that you can use it as your private RabbitMQ cluster on the cloud. Alternatively, as we will see in the next recipe, it's possible to access the RabbitMQ cluster from the Internet through a public load balancer.

See also

Read the *VPC Whitepaper* and *Security Whitepaper* PDF documents from `http://aws.amazon.com/vpc/`.

Using AWS Load Balancing in front of a RabbitMQ cluster

In the previous recipe we have created a RabbitMQ cluster. Now we will see how to use the cluster from the Internet. A typical scenario is to create a public RabbitMQ as a service.

We use Amazon Elastic Load Balancing (`http://aws.amazon.com/elasticloadbalancing/`) as the frontend.

Getting ready

You need to create a cluster as we have already seen in the *Creating a cluster with two EC2 instances* recipe.

How to do it...

After creating a private cluster, you can't access it directly: you need to use a gateway; let's see the AWS elastic load balance configuration:

1. Open the EC2 console and select **Load Balancers**, and on the left pane click on **Create a load balancer**.
2. Create a load balancer for your VPC.

3. Choose the ports to redirect to:
   - ❑ 5672 (RabbitMQ broker port)
   - ❑ 15672 (Web management port)
4. Configure the health check using the 5672 port.
5. Configure a new security group to open the two ports on the Internet.
6. Wait for a few minutes for the creation of the load balancer.
7. Use the load balance URL.

How it works...

The AWS load balancer works like other balancers, but you don't need to set up the software.

The load-balancer wizard is easy to use; you need to select what you want to balance, which in our case is the VPC:

Then, for port mapping, we have mapped the AMQP and the management port of the instances to the same ports of the load balancer, as you can see in the following screenshot:

Load Balancer Protocol	Load Balancer Port	Instance Protocol	Instance Port
HTTP	15672	HTTP	15672
TCP	5672	TCP	5672

The load balancer needs a health check, so we use the RabbitMQ port as shown in the following screenshot:

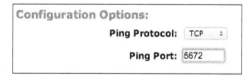

 When left idle, the AWS load balancer can eventually stop forwarding the requests after a timeout. As soon as a client begins to perform requests through the load balancer, it is not immediately reactive, but it will be activated in a few seconds.

The last step is to configure the load-balancer security group, to decide which ports are enabled:

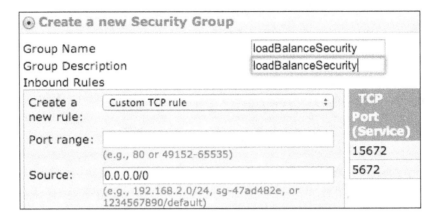

We have chosen a new security group to divide the security responsibility. This is better than using the default security group and adding the ports to it. In this way, you can manage accessibility rules with maximum flexibility.

Well, your AWS elastic load balancer is ready; you just need to wait for a few minutes to use it. Check its status by navigating to **EC2 console** | **Load Balancer**. At last you should have an output similar to the following screenshot:

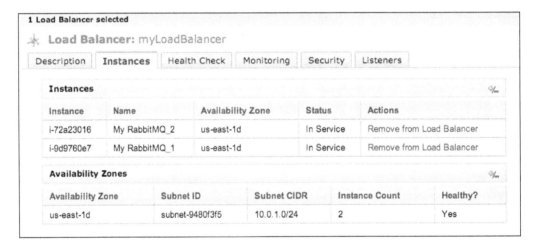

In order to work correctly, the instance status must be **In Service**, otherwise the instances are removed automatically from the balancer.

In the **Description** tab, you can find the public URL; use it to browse the console or to connect your AMQP clients as shown in the following screenshot:

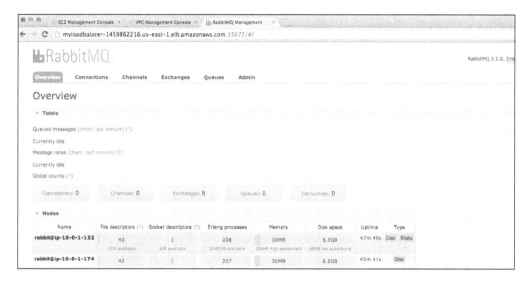

There's more...

The load balancer is usually used with AmazonRoute53 (`http://aws.amazon.com/route53/`) to map your balancer with a human-readable DNS name.

If you don't want to use the AWS elastic load balancer, you can install another EC2 instance onto the VPC public network and install your own load balancer, for example, HA Proxy or Crossroads (as we saw in the *Using a load balancer in front of consumers* recipe in *Chapter 6, Developing Scalable Applications*). Then you can add an Elastic IP to your EC2 instance.

What to use depends on the needed scalability of your cluster and which **SLA (Service Level Agreement)** you have to guarantee to your customers.

See also

With the AWS balancer you can create an internal balancer. This is an interesting feature if you have a public service that needs to communicate with your internal RabbitMQ cluster. Visit `http://aws.amazon.com/elasticloadbalancing/` and read the *Features of Elastic Load Balancing* section for more details.

Configuring EC2 dynamic bind

One of the benefits of the cloud is the scalability, and when you configure a RabbitMQ cluster, you need to automate the joining steps. You shouldn't manually configure each new RabbitMQ cluster node.

In this recipe we will see how to join new machines to the RabbitMQ cluster.

Getting ready

You need to create a VPC as we did in the *Creating a cluster with two EC2 instances* recipe.

How to do it...

In this example we use Ubuntu 12.04.3 LTS Cloud. Perform the following steps to join new machines to the cluster:

1. Create a VPC as we did in the *Creating a cluster with two EC2 instances* recipe, without creating the cluster.
2. Add one machine to the VPC. This machine will be the master of the RabbitMQ cluster, to which other nodes will be added.
3. Create the script to join a new machine to the master node of the cluster.
4. Put the script on the user data EC2 instance during the startup.
5. Check your cluster.

How it works...

After step 1, we are ready to add the machines to the cluster. We need to add the first machine from the saved snapshot created in the *Creating a master image* recipe.

The machine will be the master (in our case with 10.0.1.132 as the internal IP).

Now the master is ready; so we can create a Bash script to join the new machine to the cluster using the following code:

```bash
#!/bin/bash
/etc/init.d/rabbitmq-server stop >> /var/log/rabbitmqstartup.log
/etc/init.d/rabbitmq-server start >> /var/log/rabbitmqstartup.log
rabbitmqctl stop_app>> /var/log/rabbitmqstartup.log
rabbitmqctl reset >> /var/log/rabbitmqstartup.log
abbitmqctl join_cluster rabbit@ip-10-0-1-132 >>
  /var/log/rabbitmqstartup.log
rabbitmqctl start_app>> /var/log/rabbitmqstartup.log
rabbitmqctl cluster_status>> /var/log/rabbitmqstartup.log
```

 You can find this script in the book archive, in the directory Chapter10/ Recipe05.

This script creates a log file in `/var/log/rabbitmqstartup.log`, which is useful to check the execution of all the operations in case something goes wrong.

Now, we can add a second machine always using the master image as already seen; however, while running the wizard we can insert the script mentioned previously in the user data field to customize the instance at its first boot as shown in the following screenshot:

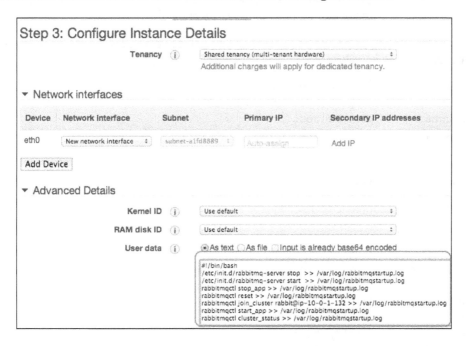

The new machine will join the cluster automatically as soon as it is set in running state.

If you want to add more than one machine at the same time, modify the **Number of Instances** value in the wizard, as shown in the following screenshot:

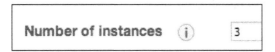

Well, if you add a load balancer (as we have seen in *Using AWS Load Balancing in front of a RabbitMQ cluster*), you can add new machines to your cluster with zero impact on the clients connected to the balancer.

There's more...

Using a single master node to join new nodes could be a single point of failure; if the master is down, no new machines can join the cluster.

An internal load balancer can avoid this problem. We can query the balancer to retrieve the available nodes, and use one of them to join the new ones.

For example, with a Python (version 2.7.x) script, we can call the /nodes web API to retrieve the available nodes, as follows:

```
url = 'http://internalloadbalancer:15672/api/nodes'
print prefix + 'Get json info from ..' + url
request = urllib2.Request(url)
base64_string = base64.encodestring('%s:%s' % ('guest',
   'guest')).replace('\n', '')
request.add_header("Authorization", "Basic %s" % base64_string)
data = json.load(urllib2.urlopen(request))
```

Then, check the first running node and join it as follows:

```
for r in data:
if r.get('running'):
print(prefix + 'found running node to bind..')
fromsubprocess import call
call(["rabbitmqctl", "stop_app"])
call(["rabbitmqctl", "reset"])
call(["rabbitmqctl", "join_cluster",r.get('name')])
call(["rabbitmqctl", "start_app"])
break
pass
```

You can replace the Bash script with this Python script. For a load balancer, you can always use the AWS load balancer by using the internal flag, as shown in the following screenshot:

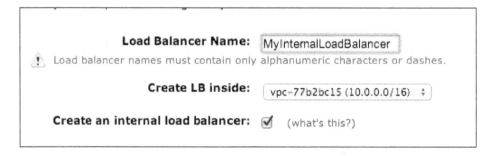

You can find the Python script at Chapter10/Recipe05/rabbitmq_startup.py.

After joining to the cluster, you should add the new machines to the load balancer, using the command-line AWS tool elb-register-instances-with-lb. Find more details of this at `http://docs.aws.amazon.com/ElasticLoadBalancing/latest/DeveloperGuide/US_DeReg_Reg_Instances.html`.

See also

To launch a new machine you can also use the command-line tool by executing the following command:

```
ec2-run-instances --key KEYPAIR --user-data-file
  clusterjoinscriptami-xxxx
```

Check `http://aws.amazon.com/cli/` for more information about AWS Command Line Interface and `http://docs.aws.amazon.com/AWSEC2/latest/CommandLineReference/ApiReference-cmd-RunInstances.html` for more information about the ec2-run-instances.

Dealing with load spikes and resource optimization in the cloud

Until now we have seen how to scale a RabbitMQ cluster manually, but in some cases it's necessary to automate this process. A typical use case is to manage load spikes automatically.

In this recipe we will see how to create an auto-scaling RabbitMQ cluster using AWS auto-scaling (`http://aws.amazon.com/autoscaling/`). Currently, AWS auto-scaling is only available using AWS command-line tools. In this recipe we have used AWS Java command-line tools (`http://aws.amazon.com/developertools/351`).

 The auto-scaling web interface has become available after the chapter has been written. You can find the documentation at `http://docs.aws.amazon.com/AutoScaling/latest/DeveloperGuide/GettingStartedTutorial.html`.

Getting ready

To run this recipe you will need to create a VPC, as we have seen in the *Creating a cluster with two EC2 instances* recipe, avoiding adding any machine to it. Then, you need an EC2 master image similar to the one created in the *Creating a master image* recipe.

In this example you need to install Python 2.7.x (already present if you are using Ubuntu 12.04.3 LTS Cloud).

How to do it...

As in the previous recipes, we use Ubuntu 12.04.3 LTS Cloud. Perform the following steps to create an auto-scaling RabbitMQ cluster:

1. Create a VPC as we did in the *Creating a cluster with two EC2 instances* recipe, without creating the cluster.

2. Create an internal AWS load balancer as we have seen in the *Configuring EC2 dynamic bind* recipe.

3. Create a Python script to join the new machine to the cluster using the following code:

```
fromsubprocess import call
call(["rabbitmqctl", "stop_app"])
call(["rabbitmqctl", "reset"])
try:
url = 'http://yourinternallb:15672/api/nodes'

request = urllib2.Request(url)
```

4. You can find the full source code from the book code repository in the book repository archive at `Chapter10/Recipe6`.

5. Store the script on AWS S3. Open the S3 console and upload the script file, for example, `https://s3.amazonaws.com/rabbitmq_startup/rabbitmq_startup.py` as shown in the following screenshot:

6. Create a new master machine and insert the following commands in the boot script `/etc/rc.local`:

```
wget
  https://s3.amazonaws.com/rabbitmq_startup/rabbitmq_startup.py
    -O /tmp/rabbitmq_startup.py

sudo python /tmp/rabbitmq_startup.py
```

Then, create a new master image.

7. Choose a machine to manage the auto-scaling; you can use your local PC or an AMI. We suggest you use an AMI machine, for example, the NAT machine.

8. Configure a launch group by executing the following command:

```
as-create-launch-configRabbitMQLC --image-id yourmasteramid --
    instance-type m1.small --region us-east-1
```

9. Configure an auto-scaling group by executing the following command:

```
as-create-auto-scaling-group RabbitMQGroup --launch-
    configuration RabbitMQLC --region us-east-1 --availability-
        zones us-east-1e --min-size 1 --max-size 20 -vpc-zone-
            identifier "yourprivatesubnet" --load-balancers
                myloadbalancerName  --health-check-type ELB --grace-
                    period 600
```

10. Configure a scaling-policy to let the cluster grow by executing the following command:

```
as-put-scaling-policyRabbitMQScaleUpPolicy --auto-scaling-
    group RabbitMQGroup --adjustment=1 --region us-east-1 --type
        ChangeInCapacity
```

11. Configure a scaling-policy to let the cluster shrink by executing the following command:

```
as-put-scaling-policyRabbitMQScaleDownPolicy --auto-scaling-
    group RabbitMQGroup --adjustment=-1 --region us-east-1 --
        type ChangeInCapacity
```

12. Configure the Cloud Watch alarm, opening the Cloud-Watch console and define an alarm, for example, on the CPU usage, which triggers a Scaling-up policy as shown in the following screenshot:

13. Okay, the cluster should be ready with just one machine up and running.

14. Reset auto-scaling by executing the following command:

```
as-update-auto-scaling-group RabbitMQGroup --min-size 0 --max-
    size 0
```

15. Remove auto-scaling by executing the following command:

```
as-delete-auto-scaling-group RabbitMQGroup
```

```
as-delete-launch-config RabbitMQLC
```

How it works...

The aim of this recipe is to create a RabbitMQ cluster that scales automatically without manual operations using AWS auto-scaling: when the cluster needs more resources, the configured AWS services will trigger the addition of a new instance, created from the configured AMI.

Once the new instance has started, it's important that we provide all the instructions for the instance to:

- ▶ Join the existing private subnet of the VPC
- ▶ Join the existing load balancer
- ▶ Join the existing RabbitMQ cluster

For the last step, we need a boot Python script that queries the load balancer to retrieve one RabbitMQ node and join the machine.

We decided to put such scripts outside the AMI, because if you want to change something you don't need to recreate it, but just publish a new script version.

 Inside the script that you find in the book sources, remember to change the internal load-balancer URL, and in boot startup, remember to change the S3 URL.

So, now all the pieces are ready, and we can configure the AWS auto-scaling. Currently, there isn't a web interface to configure the auto-scaling: you can configure it using AWS Java command-line tools.

 The VPC wizard creates an AMI NAT machine, so you can use it where the command-line tools are already installed and you will just need to configure the access credentials.

By following step 7 you create a launch configuration, and with step 8 an auto-scaling group.

 We are using the Java **command line interface** (**CLI**), but you can also use the Python CLI. Check this link for more details: `http://aws.amazon.com/cli/`

In this configuration we decide to launch an `m1.small` instance from the master AMI. The group can have minimum one machine and maximum 20 machines and will be deployed to your private VPC-zone; then, the machine will be attached to the internal load balancer.

When you define the scaling group, it will automatically start one machine, and now you should have two running machines, one is the NAT machine and one is the first cluster machine as shown:

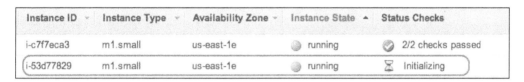

Instance ID	Instance Type	Availability Zone	Instance State	Status Checks
i-c7f7eca3	m1.small	us-east-1e	running	2/2 checks passed
i-53d77829	m1.small	us-east-1e	running	Initializing

Now we need to create the auto-scaling policies, one to grow the cluster, and one to shrink it (steps 9-10).

 You can try the policy using:

`as-execute-policy RabbitMQScaleUpPolicy`

Now there will be machines!

The last step to activate the automatic scaling is to create a Cloud-Watch alarm, as shown in step 11, and you can choose different metrics. In our simple case we have chosen a CPU alarm.

When an alarm is raised, a new machine will be launched using the `RabbitMQScaleUpPolicy` command, as shown in the following screenshot:

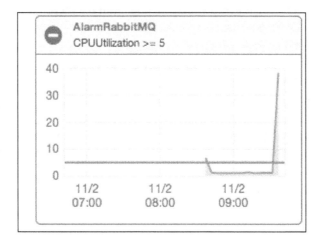

 CPU utilization >= 5 percent is a very low value; use it only for testing purposes.

The cluster can grow up to 20 machines. In the example we have performed a stress test to raise the alarm (you can find the source code at `Chapter10/Recipe6_StressTest`). Once the JAR file has been created using:

mvn clean compile assembly:single

You can copy the JAR file to one machine on the public subnet, and then execute the following command for the producer:

**java -cp rmqStressTest.jar rmqexample.Producer YOURinternal-
    loadbalancer**

For the consumer, execute the following command:

java -cp rmqStressTest.jar rmqexample.Consumer YOURinternalloadbalancer

You will see the cluster growing quickly as shown in the following screenshot:

	Name	aws:autoscaling:gr	Instance ID	Instance Type	Availability Zone	Instance State
			i-c7f7eca3	m1.small	us-east-1e	running
		RabbitMQGroup	i-3a3ce05c	m1.small	us-east-1e	running
		RabbitMQGroup	i-cfa5f6aa	m1.small	us-east-1e	running
		RabbitMQGroup	i-2da5f648	m1.small	us-east-1e	running
		RabbitMQGroup	i-35d7424c	m1.small	us-east-1e	running
		RabbitMQGroup	i-9634c7eb	m1.small	us-east-1e	running
		RabbitMQGroup	i-304db357	m1.small	us-east-1e	running
		RabbitMQGroup	i-3b04b45c	m1.small	us-east-1e	running
		RabbitMQGroup	i-bd5751c7	m1.small	us-east-1e	running

Filter: All instances ˅   All instance types ˅   Q Search Instances

Remember, the cluster is inside a VPC; so to connect to it you have to use the NAT machine as we have seen in the *Creating a cluster with two EC2 instances* recipe. Once connected to one cluster machine, you can execute:

```
rabbitmqctl cluster_status
```

If everything is ok, you should have the result shown in the following screenshot:

```
Cluster status of node 'rabbit@ip-10-0-1-115' ...
[{nodes,[{disc,['rabbit@ip-10-0-1-103','rabbit@ip-10-0-1-104',
                'rabbit@ip-10-0-1-115','rabbit@ip-10-0-1-117',
                'rabbit@ip-10-0-1-119','rabbit@ip-10-0-1-150',
                'rabbit@ip-10-0-1-24','rabbit@ip-10-0-1-95']}]},
 {running_nodes,['rabbit@ip-10-0-1-103','rabbit@ip-10-0-1-24',
                'rabbit@ip-10-0-1-150','rabbit@ip-10-0-1-117',
                'rabbit@ip-10-0-1-95','rabbit@ip-10-0-1-119',
                'rabbit@ip-10-0-1-104','rabbit@ip-10-0-1-115']},
 {partitions,[]}]
...done.
```

Okay, now your cluster can manage load spikes.

In order to remove auto-scaling, you have to execute the following commands:

```
as-update-auto-scaling-groupRabbitMQGroup --min-size 0 --max-size 0
as-delete-auto-scaling-groupRabbitMQGroup
as-delete-launch-config RabbitMQLC
```

If you get an error during the deletion of auto-scaling group, manually remove the cluster machines and retry.

 If you just delete the machines, auto-scaling will restart them automatically—we have configured it for this purpose! To remove auto-scaling, you must execute the delete procedure and drop it first.

There's more...

Building an auto-scaling application is very hard. With this recipe you should have the basic knowledge to create an auto-scaling RabbitMQ cluster. In this recipe, you can see how to increase the number of cluster machines, but it's also important to know how to decrease them.

You can remove a machine by just executing the `RabbitMQScaleDownPolicy` command. However, removing a machine from the cluster is not that easy, because you must be careful in case the node has queues with stored messages. In this case you can adopt some tips, for example, creating a pair mirror queue as we saw in the *Optimizing mirror policies* recipe in *Chapter 7, Developing High-availability Applications*.

When it's time to remove a machine, don't forget to remove it from the RabbitMQ cluster in advance.

In this recipe we have seen how to create a cluster within a single AWS availability zone. In order to create a geographically distributed cluster, you should create two or more clusters on different availability of zones and synchronize them using Federation or Shovel as we saw in *Chapter 7, Developing High-availability Applications*.

See also

Monitoring a cluster can be very hard, especially when your cluster is dynamic. It's possible to use the existing monitoring tools that we will discuss in the next chapter.

11
AMQP and Cloud Computing – RabbitMQ on PaaS

In this chapter we will cover:

- ▸ RabbitMQ on CloudAMQP
- ▸ First application on Cloud Foundry
- ▸ Using RabbitMQ on Cloud Foundry

Introduction

In this chapter we will talk about the **PaaS** (**Platform as a service**) cloud and some cloud service providers that have RabbitMQ among the available services.

PaaS commonly refers to a cloud computing service model, where not only the physical servers and the networking infrastructure, but also the software stack is handled by the service provider, relieving the customer from administration efforts.

In this chapter we will learn about some vendors that provide RabbitMQ as a ready-to-use software stack.

RabbitMQ on CloudAMQP

The www.cloudamqp.com website is not exactly a PaaS cloud platform, but we need to introduce it to understand the next recipe.

CloudAMQP is a RabbitMQ service. You can use this service to avoid management and hardware costs. There are five plans available at `http://www.cloudamqp.com/plans.html` that you can choose from. The **Little Lemur** plan is free and is perfect for trying this service.

Getting ready

You require the following software:

▶ Java 1.6+

▶ Apache Maven

How to do it...

To try this recipe, you need to register an account on CloudAMQP; the registration is free.

1. Register an account at `https://customer.cloudamqp.com/login`.
2. Create a RabbitMQ instance; choose the **Data center** and the **Plan**:

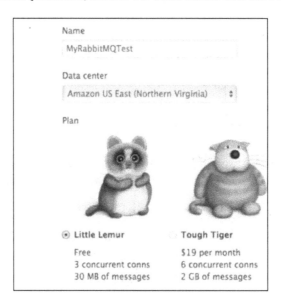

3. Go to the instance's details to see all your RabbitMQ connection parameters.
4. Create a Java Maven program and add the following dependency:

```
<dependency>
<groupId>com.rabbitmq</groupId>
<artifactId>amqp-client</artifactId>
<version>2.8.1</version>
</dependency>
```

5. Create a Java `Producer` class that creates and uses the queue called `myorders_11`.

6. Create a Java `Consumer` class that creates and uses the queue called `myorders_11`.

You can find the source code at `Chapter11/Recipe01/`.

How it works...

After creating your account (step 1), you can start using a RabbitMQ instance.

CloudAMQP service, by default, runs on Amazon AWS, and you have to choose the AWS data center (as we have seen in the previous chapter).

Go to the **Instance Details** page to see your connection parameters:

In the source code example, we had created a producer and a consumer, bound to the queue `myorders_11`. The example sends a simple message to the queue, using the producer, and receives the message using the consumer.

> To test this example, you can pass the CloudAMQP URL as a parameter in the following manner:
>
> ```
> java -cp   rabbitmq-client.jar:Recipe01-0.0.1-
> SNAPSHOT.jar rmqexample.Consumer amqp://oxxx:xxxt-
> H74UosCJiN_xxx_jE@lemur.cloudamqp.com/oibyfhxg
> ```

There's more...

CloudAMQP is a very complete RabbitMQ service cloud. You can find all the clients and other cloud integrations at `http://www.cloudamqp.com/docs.html`.

First application on Cloud Foundry

The open PaaS, `www.cloudfoundry.com`, has been recently updated to Version 2.0. It is still in the beta phase and is incomplete, but the definitive release shouldn't be too far off.

One of the most interesting features to use is a local virtual machine with Cloud Foundry v2.0.

Using this feature, the software development is done very quickly because you can work, test, and deploy your application locally in your lab without waiting for uploads. Go to `http://docs.cloudfoundry.com/docs/running/deploying-cf/run-local.html` for more information. In this recipe, we deploy our first **Spring** application in the cloud.

Getting ready

You will need the following software:

- ▶ Ruby and RubyGems (See `http://docs.cloudfoundry.com/docs/common/install_ruby.html` for installation details)
- ▶ Java 1.6+
- ▶ Apache Maven
- ▶ Spring Framework (we suggest that you use the Spring Tool suite)

How to do it...

You can deploy your first application by following these ten steps:

1. Register an account on Cloud Foundry by visiting `https://console.run.pivotal.io/register`.

2. Install the `cf` command line tool using the following command:

   `gem install cf`

3. Choose the default target using the following command:

   `cf target api.run.pivotal.io`

4. Log in to the cloud using the following command:

   `cf login`

5. Choose one of the following workspaces:

   - ❑ **Development**
   - ❑ **Production**
   - ❑ **Staging**

6. Create your Spring MVC application, or alternatively, just use the example located in `Chapter11/Recipe02` directly.

7. Make your application using the following command:

   `mvn package`

8. Publish your application using the following command:

```
cf push --path target/rmq-1.0.1-BUILD-SNAPSHOT.war
```

9. Follow the `cf` steps.

10. Check the status on the console.

How it works...

After installing the Cloud Foundry command line tool (`cf`), you need to choose the deployment target, which can be a local virtual machine or the Cloud Foundry system. In this recipe, we use the latter by executing the following command:

```
cf target api.run.pivotal.io
```

Now, execute the login with the command line tool using the command:

```
cf login
```

Insert your username and password, and then choose your workspace as shown in the following screenshot:

```
Gabrieles-MacBook-Pro:Recipe02 gabriele$ cf login
target: https://api.run.pivotal.io

Email>      ------·-------- o@gmail.com

Password> ********

Authenticating... OK
1: development
2: production
3: staging
Space> 1

Switching to space development... OK
```

You are now logged onto the cloud and can deploy your application.

To start quickly, you can use the source code at `Chapter11/Recipe02` and execute the following command on the root folder to create the `war` archive:

```
mvn package
```

You are now ready to deploy your application on the cloud. Execute the following command in the same directory:

```
cf push --path target/rmq-1.0.1-BUILD-SNAPSHOT.war
```

When you execute the `cf` push command you have to choose the publishing parameters, as shown in the following screenshot:

```
Gabrieles-MacBook-Pro:Recipe02 gabriele$ cf push --path target/rmq-1.0.1-BUILD-SNAPSHOT.war
Name> myfirstapplication

Instances> 1

1: 128M
2: 256M
3: 512M
4: 1G
Memory Limit> 256M

Creating myfirstapplication... OK

1: myfirstapplication
2: none
Subdomain> myfirstapplication

1: cfapps.io
2: none
Domain> cfapps.io

Binding myfirstapplication.cfapps.io to myfirstapplication... OK
```

You can answer the other command questions with the default values, and the application is deployed!

When the deployment is terminated, you will see the following screenshot:

```
Checking status of app 'myfirstapplication'........
   0 of 1 instances running (1 starting)
   0 of 1 instances running (1 starting)
   0 of 1 instances running (1 starting)
   0 of 1 instances running (1 starting)
   0 of 1 instances running (1 starting)
   1 of 1 instances running (1 running)
Push successful! App 'myfirstapplication' available at http://myfirstapplication.cfapps.io
```

While checking the Cloud Foundry console, you will see the following:

You can finally go to the URL `http://myfirstapplication.cfapps.io` to actually use your first application that has been deployed on Cloud Foundry.

The application name is unique on the cloud, so if `myfirstapplication` is already in use, please choose another one.

There's more...

Cloud Foundry is an open PaaS and the documentation is very precise. See `http://docs.cloudfoundry.com` for more information and `http://docs.cloudfoundry.com/docs/dotcom/getting-started.html#push-app` for the complete reference.

Using RabbitMQ on Cloud Foundry

Cloud Foundry uses CloudAMQP (see the *RabbitMQ on CloudAMQP* recipe) as a RabbitMQ service. In the previous recipes we have introduced the two cloud services. Now we will show how to use RabbitMQ on Cloud Foundry.

You can find the source at `Chapter11/Recipe03/`.

Getting ready

We will need the following software:

- ▸ Ruby
- ▸ RubyGems
- ▸ Java 1.6+
- ▸ Spring Framework
- ▸ Apache Maven

How to do it...

For a quick startup, we have copied the source code from the recipe *RabbitMQ on CloudAMQP* and added the RabbitMQ configuration.

1. Log in to the Cloud Foundry system and open the section **SERVICES IN DEVELOPMENT**, then add a new CloudAMQP service. Finally, choose the free instance as shown in the following screenshot:

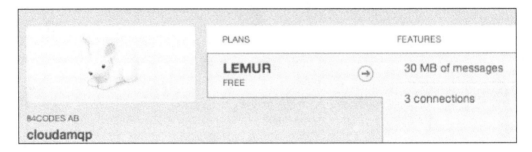

2. Call the CloudAMQP service `amqp1`.

3. Open the `POM.xml` Maven file and add the Spring AMQP reference as shown in the following code:

```
<dependency>
<groupId>org.springframework.amqp</groupId>
<artifactId>spring-rabbit</artifactId>
<version>1.2.0.RELEASE</version>
</dependency>
```

4. Also add the `Jackson` library (`http://jackson.codehaus.org/`):

```
<dependency>
<groupId>org.codehaus.jackson</groupId>
<artifactId>jackson-mapper-asl</artifactId>
<version>1.9.2</version>
</dependency>
```

5. Open the `servlet-context.xml` file; then add RabbitMQ and a Cloud Foundry schema:

```
xmlns:rabbit="http://www.springframework.org/schema/rabbit"
xmlns:cloud="http://schema.cloudfoundry.org/spring"
```

6. Also, add the following RabbitMQ and Cloud Foundry schema locations:

   ❑ `http://schema.cloudfoundry.org/spring`
   ❑ `http://schema.cloudfoundry.org/spring/cloudfoundry-spring-0.8.xsd`

- http://www.springframework.org/schema/rabbit
- http://www.springframework.org/schema/rabbit/spring-rabbit-1.1.xsd

7. Add the RabbitMQ configuration to the `servlet-context.xml` file:

```
<cloud:rabbit-connection-factory service-name="amqp1"
id="connectionFactory" />

<rabbit:template id="rabbitTemplate" connection-
factory="connectionFactory" />

<rabbit:admin connection-factory="connectionFactory"/>
```

8. Add the queue name as shown in the following code:

```
<rabbit:queue name="myorders_11"durable="true"/>
```

9. Create an HTTP handler that is able to manage the HTTP requests with the `/rest` URI. The handler is a Java class with `@Controller` and `@RequestMapping` Spring annotations:

```
@Controller
@RequestMapping("rest")
public classBooksController{…}
```

10. Add two methods to the `BooksController` class. The first one is for getting the book list:

```
@ResponseBody public Collection<Book>getBooks(){…
}
```

The second one is for managing the book purchase workflow:

```
public @ResponseBody String buybook(@PathVariable("bookid")
intbookid){....
}
```

11. See http://static.springsource.org/spring/docs/3.2.x/spring-framework-reference/html/mvc.html#mvc-introduction for more details about Spring MVC and spring controller.

12. The server is ready. Now, on the client side, add two JavaScript methods, the first one, `$.getJSON('/rest/books', function(data)..)`, to get the book list and the second one to execute the call `buyBook`:

```
$.ajax({
type: "GET",
url: "rest/buybook/"+bookid,
error: function(xhr, status, error) {
```

```
$("#labelinfo").text("Error request: " +error);

    },
      success: function(xhr, status, error) {
      $("#labelinfo").text(xhr);


    }
  });
```

13. The application is ready. Now, deploy the application on the cloud using the `cf` tool (as we have seen in the recipe *First application on Cloud Foundry*), and bind the **CloudAMQP** service (steps 1 and 2) to the application as shown in the following screenshot:

```
Gabrieles-MacBook-Pro:Recipe03 gabriele$ cf push --path ta
Name> myrabbitmqtest

Instances> 1

1: 128M
2: 256M
3: 512M
4: 1G
Memory Limit> 3

Creating myrabbitmqtest... OK

1: myrabbitmqtest
2: none
Subdomain> myrabbitmqtest

1: cfapps.io
2: none
Domain> cfapps.io

Creating route myrabbitmqtest.cfapps.io... OK
Binding myrabbitmqtest.cfapps.io to myrabbitmqtest... OK

Create services for application?> n

Bind other services to application?> y

1: jm
2: amqp1
Which service?> 2

Binding amqp1 to myrabbitmqtest... OK
```

14. Execute the consumer from the *RabbitMQ on CloudAMQP* recipe using the CloudAMQP-RabbitMQ URL created by the service `amqp1` (step 2).

How it works...

In this example, we have created a minimal real-use case using Spring Framework. We have simulated a book store, where the user buys a book, the server stores the request into the cloud database, and then sends a message with book details to `myorders_11` queue. An external consumer gets the message and redirects the order to the company store. Let's see this in detail.

First of all, we create a RabbitMQ instance called `amqp1` (step 1). Cloud Foundry uses CloudAMQP as a RabbitMQ cloud service (you can see the recipe *RabbitMQ on CloudAMQP*) and the `amqp1` service is accessible outside via the cloud.

Also, in this example we use Spring with Maven, and in order to use the RabbitMQ instance, we need to add the Spring AMQP package to the Maven `POM.xml` file(step 3).

We have used **Jackson** to create a **REST** application (`http://en.wikipedia.org/wiki/Representational_state_transfer`).

> The book's REST URL is `http://myrabbitmqtest.cfapps.io/rest/books`

With steps 5 and 6, we define the schema and the schema location for Cloud Foundry and RabbitMQ. We can configure the connection factory with these two schemas.

> If you have more than one CloudAMQP service, you must set the service name; in our case it is `amqp1`. (See `http://docs.cloudfoundry.com/docs/using/services/spring-service-bindings.html#rabbitmq` for more details)

In the same file, we can define the queue name using:

```
<rabbit:queue name="myorders_11" durable="true"/>
```

The configuration is terminated, and in the server, we implement two methods (steps 9 and 10): `getbooks` and `buybook`.

The first one gets the list of books as a JSON string (`/rest/books`) and the second one gets a book ID from the client and sends a message to the `myorders_11` queue:

```
public @ResponseBody String buybook(@PathVariable("bookid") intbookid)
...
Book bbook = myBooks.get(bookid);
String message = mapper.writeValueAsString(bbook);
amqpTemplate.convertAndSend("myorders_11",message);
```

The `amqpTemplate` variable is a Spring `@Autowired` annotation that uses the RabbitMQ parameters defined to the file `servlet-context.xml`(steps 5 and 7).

See `http://static.springsource.org/spring/docs/3.2.x/spring-framework-reference/html/beans.html#beans-autowired-annotation` for more information about `@Autowired` annotations.

From the client side, executing one JavaScript request for the books and another for buying the book is sufficient (step 11).

The website is now ready to be deployed; we just need to follow the steps from the recipe *First application on Cloud Foundry* and link our RabbitMQ instance to the application (step 11).

> The `amqp1` service can be shared with different applications.

After deploying the application, you can finally see the website online, and when you execute a consumer from your local machine, you will see the book orders as shown in the following screenshot:

As a consumer, you can use the one from the *RabbitMQ on CloudAMQP* recipe. You just need to change the CloudAMQP parameters in the `amqp1` service and it's done! We have just created a frontend site that sends book orders to our local machine/machines in real time.

There's more...

You can use the web interface to check the application's health and the `cf` tool to check the logs:

- ▸ `cf logs application_name` (for example, `myrabbitmqtest`) to see the logs
- ▸ `cf tail application_name` to see the log tail
- ▸ `cf crashlogs application_name` to retrieve the crash logs

There are other ways to deploy and manage the application on Cloud Foundry, for example, by using the Cloud Foundry Eclipse plugin (`http://marketplace.eclipse.org/content/cloud-foundry-integration-eclipse`). After logging into Cloud Foundry, you can easily access the features.

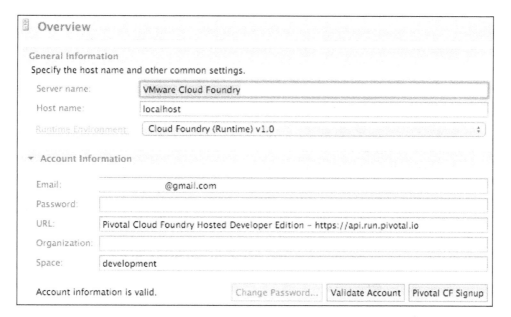

You can check your application and services, and directly add your services in the **Applications** window, as shown in the following screenshot:

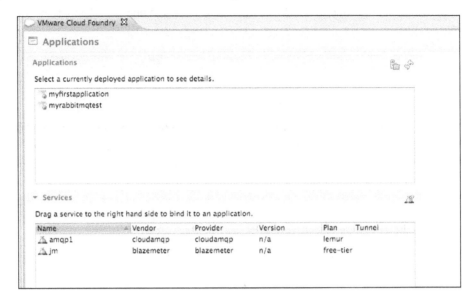

It is also possible to check the application logs on the **Remote Server** eclipse view, as shown in the following screenshot:

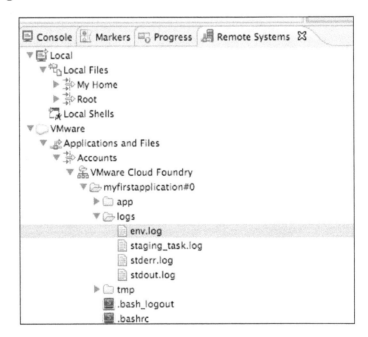

12
Managing RabbitMQ Error Conditions

In this chapter, we will cover the following topics:

- ▶ Monitoring RabbitMQ's behavior
- ▶ Using RabbitMQ to troubleshoot itself
- ▶ Tracing RabbitMQ's ongoing activity
- ▶ Debugging RabbitMQ's messages
- ▶ What to do when RabbitMQ fails to restart
- ▶ Debugging using Wireshark

Introduction

Whenever we develop an application, it's a common practice to develop a diagnostic infrastructure. This can be based on log files, SNMP traps, and many others.

RabbitMQ provides both standard log files and a built-in messaging-based troubleshooting solution.

We will see how to use these features in the first three recipes.

Sometimes, there are problems that prevent RabbitMQ from starting. In this case, it's mandatory to fix the problem directly on the server machine where the issue persists, and to reset the broker. We'll see this in the *What to do when RabbitMQ fails to restart* recipe.

However, debugging messages is a part of application development too. In this case, we need to know the exact information exchanged between RabbitMQ and its clients. It is possible to use a proxy built-in tool, part of the Java client API (see the *Debugging RabbitMQ's messages* recipe) or to use an advanced network monitor to examine the traffic, as we will see in the *Debugging using Wireshark* recipe.

Monitoring RabbitMQ's behavior

In order to check the correct behavior of RabbitMQ, it is useful to have a monitoring tool, especially when dealing with a cluster.

There are many different tools, commercial and freeware, which help to keep things under control in distributed systems such as Nagios and Zabbix.

In this recipe, we will show how to configure the RabbitMQ plugin for **Ganglia** (`http://sourceforge.net/apps/trac/ganglia/wiki/ganglia_quick_start`).

Getting ready

In order to run this recipe, you need to have RabbitMQ configured with the management plugin enabled.

You need to install and configure Ganglia too. In this recipe, we have used version 3.6.0.

How to do it...

In order to see RabbitMQ statistics within Ganglia monitoring graphs, you need to perform the following steps:

1. Install and configure Ganglia using `yum` or `apt-get` depending on your Linux distribution. You will need the following packages:
   - `ganglia-gmetad`
   - `ganglia-gmond`
   - `ganglia-gmond-python`
   - `ganglia-web`
   - `ganglia`

2. Copy the Python Ganglia monitor plugin into `/usr/lib64/ganglia/python_modules` from `https://github.com/ganglia/gmond_python_modules/blob/master/rabbit/python_modules/rabbitmq.py`.

3. Copy the Python Ganglia configuration file into `/etc/ganglia/conf.d` from `https://github.com/ganglia/gmond_python_modules/blob/master/rabbit/conf.d/rabbitmq.pyconf`.

4. Check the configuration file for the correct parameters. In particular, you probably need to fix the following entry by leaving only the default vhost:

```
paramvhost {
value = "/"
    }
```

5. If it is already running, restart gmond using:

```
service gmond restart
```

How it works...

By going through this recipe, you will be able to monitor RabbitMQ from within the Ganglia environment.

 For basic troubleshooting, you can have a look at the log files that, by default, are saved into /var/log/rabbitmq.

Once it's up and running, you will be able to access both system-wide information and information about RabbitMQ queues and nodes from the same web interface, as you can see in the following screenshot:

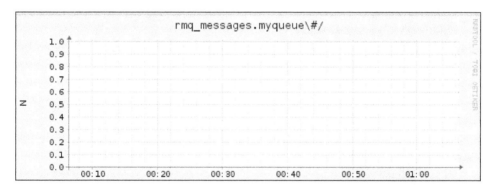

There's more...

Ganglia is a widespread solution for cluster monitoring, but not the only one.

Other typical solutions with RabbitMQ configurations that are available are Nagios (www. nagios.org), Zabbix (www.zabbix.com), and Puppet (puppetlabs.com).

Using RabbitMQ to troubleshoot itself

As mentioned in the previous recipe, we can monitor the RabbitMQ behavior by accessing its log files in quite a conventional way.

It is also possible to access the same kind of information using RabbitMQ itself, by informing a generic AMQP client of the broker activity.

Getting ready

To run this recipe, we need RabbitMQ up and running and the Java client library.

How to do it...

To consume log messages, you can execute the Java main function in `Consumer.java`. You can find this in the book source archive in the directory `Chapter12/Recipe02/Java/src/rmqexample`. In the following, we highlight the main steps:

1. Create a temporary-anonymous queue and bind it to the AMQP log exchange:

    ```
    String tmpQueue = channel.queueDeclare().getQueue();
    channel.queueBind(tmpQueue, "amq.rabbitmq.log","#");
    ```

2. In the consumer callback (`ActualConsumer.java`), retrieve the message and the routing key of each message and print them:

    ```
    String routingKey = envelope.getRoutingKey();
    String message = new String(body);
    System.out.println(routingKey + ": " + message);
    ```

3. At this point, you can run any RabbitMQ operation on the broker, and you will see it logged to the standard output.

How it works...

The RabbitMQ log exchange, `amq.rabbitmq.log`, is a topic exchange to which RabbitMQ itself publishes its log messages.

In our example code, we consume messages from all the topics using the # wildcard.

For example, by running another code that runs two connections to the same broker and by aborting it, we get the following output:

```
info: accepting AMQP connection <0.2737.0> (127.0.0.1:54698 ->
127.0.0.1:5672)
```

```
info: accepting AMQP connection <0.2753.0> (127.0.0.1:54699 ->
127.0.0.1:5672)
warning: closing AMQP connection <0.2737.0> (127.0.0.1:54698 ->
127.0.0.1:5672):
connection_closed_abruptly
warning: closing AMQP connection <0.2753.0> (127.0.0.1:54699 ->
127.0.0.1:5672):
connection_closed_abruptly
```

It is worth to note that the information, `info` and `warning` reported here is not a part of the messages themselves, but are the routing keys that we are printing at the beginning of each message (step 2 of the preceding steps).

 In case we just want to receive warning and error messages, we can subscribe to the corresponding topics only.

There's more...

By default, the log exchange, `amq.rabbitmq.log`, is created in the vhost /. It's possible to customize its location by defining `default_vhost` in the RabbitMQ configuration file.

Tracing RabbitMQ's ongoing activity

Sometimes, we need to trace all the messages that are being received and delivered by RabbitMQ, to analyze and debug unexpected application behavior.

RabbitMQ provides the so-called **firehose** tracing tool to have such information available.

The tracing activity can be enabled and disabled at runtime, and it should be used just for debugging since it imposes an overhead on the broker activity.

Getting ready

To run this recipe, we need RabbitMQ up and running and the Java client library.

How to do it...

RabbitMQ sends trace messages using the same mechanism used for log messages; so, the example code is very similar to the one of the previous recipe.

To consume trace messages, you can execute the Java main function in `Consumer.java` that you can find in the book source archive in the directory `Chapter12/Recipe02/Java/src/rmqexample`. Here, we highlight the main steps:

1. Create a temporary queue and bind it to the AMQP log exchange:

```
String tmpQueue = channel.queueDeclare().getQueue();
channel.queueBind(tmpQueue,"amq.rabbitmq.trace","#");
```

2. In the consumer callback (`ActualConsumer.java`), retrieve the messages with some more information for each one and print them using the following code:

```
String routingKey = envelope.getRoutingKey();
String message = new String(body);
Map<String,Object> headers = properties.getHeaders();
LongStringexchange_name = (LongString)
headers.get("exchange_name");
LongString node = (LongString) headers.get("node");
...
```

3. Activate firehose by invoking it from the root user (Linux) or on the RabbitMQ command console (Windows). Use the following command to activate firehose:

   `rabbimqctl trace_on`

4. At this point, you can start producing and sending messages to the broker, and you will see them traced to the standard output.

5. Deactivate firehose by invoking the command:

   `rabbimqctl trace_off`

How it works...

The `amq.rabbit.trace` topic exchange, by default, does not receive any messages, but once firehose is activated (step 3 of the previous steps), all the messages traveling through the broker will be copied to it by following specific rules:

▶ Messages entering the broker are published with the routing key `publish.exchange-name`, where `exchange-name` is the name of the exchange where the message was originally published to.

▶ Messages leaving the broker are published with the routing key `deliver.queue-name`, where `queue-name` is the name of the queue where the message has been originally consumed from.

▶ The body of the message is copied from the original one.

> ▶ The metadata of the original messages are inserted in the header properties of the copied message. In step 2 of the previous numbered bullets, we have seen how to retrieve the exchange name to which the message was originally delivered, but it's possible to get all the original information, that is, find all the available fields inserted into the message properties at the firehose official documentation link at `http://www.rabbitmq.com/firehose.html`.

Debugging RabbitMQ's messages

Sometimes, it is useful to just have an idea of the messages that are really traveling through a broker by just logging all of them to the standard output.

It is possible to trace those messages using a simple application provided with the RabbitMQ Java client.

Getting ready

To run this recipe, you need to have RabbitMQ up and running on the standard port `5672` and the RabbitMQ Java client library.

How to do it...

RabbitMQ includes a tracing utility in the Java client library that you can put in action by following these steps.

1. Download the latest version of the RabbitMQ Java client library from `http://www.rabbitmq.com/java-client.html`.

2. Unpack it and enter its directory.

3. Run the Java tracer by running:

   ```
   ./runjava.sh com.rabbitmq.tools.Tracer
   ```

4. Run the Java client that is to be debugged and that connects to port `5673`. For this recipe, we will use another Java tool included in the Java client library, by invoking:

   ```
   ./runjava.sh com.rabbitmq.examples.PerfTest -h amqp://
   localhost:5673 -C 1 -D 1
   ```

How it works...

The Java tracing tool is a simple AMQP proxy; by default, it listens on port `5673` and forwards all the requests to the RabbitMQ broker listening by default on `localhost` at port `5672`.

All the messages produced or consumed, as well as the AMQP operations, are all logged to a standard output.

To run the recipe at step 4 of the previous steps we have used another tool that is included in the RabbitMQ Java client library that can be used to perform a stress test with RabbitMQ.

In our case, we have just limited it to produce one message (-C 1) and consume it (-D 1).

 The tracing tool is available in the Java client API only.

There's more...

It's possible to pass some more parameters to the Java tracing program using the following code:

```
./runjava.sh com.rabbitmq.tools.Tracer listenPort connectHost
connectPort
```

In the preceding code `listenPort` refers to the port where the tracer is listening (default: `5673`), `connectHost`/`connectPort` (default: `localhost`/`5672`) are the host and the port where it connects to and forwards the requests it receives.

You can find all `PerfTest` available options using:

```
./runjava.sh com.rabbitmq.examples.PerfTest --help
```

See also

You can find the documentation of the Java tracing tool and PerfTest at `http://www.rabbitmq.com/java-tools.html`.

What to do when RabbitMQ fails to restart

Occasionally, RabbitMQ fails to restart. This can be an important issue in case the broker contains persistent data; otherwise, it's enough to reset the broker persistent state.

Getting ready

To run this recipe, you just need a test RabbitMQ broker.

 We are going to destroy all the previously defined data—avoid using a production instance.

How to do it...

To clean-up RabbitMQ, it's enough to follow these simple steps:

1. Stop RabbitMQ if it is running.
2. Locate the **Mnesia** database directory. By default, it's `/var/lib/rabbitmq/mnesia` (Linux) or `%APPDATA%\RabbitMQ\db` (Windows).
3. Delete it recursively.
4. Restart RabbitMQ.

How it works...

The Mnesia database contains all the runtime definitions of RabbitMQ: queues, exchanges, users, and so on.

By deleting it, (or renaming it in case we want to try to recover some data, or to eventually fall back in case it's possible) RabbitMQ is reset to the factory defaults; once started, it will create a new Mnesia database and initialize it with default values.

There's more...

In case the broker fails to start the first time, it is probable that there is a file permission problem in one of the system directories: either the Mnesia database directory or the log directory or some temporary, or custom directories that are specified in the configuration file.

You can find quite an exhaustive list of cases in the RabbitMQ troubleshooting page (`http://www.rabbitmq.com/troubleshooting.html`).

See also

You can find more information on how to hack Mnesia databases at the Mnesia API documentation pages (`http://www.erlang.org/doc/man/mnesia.html`).

Debugging using Wireshark

In the *Debugging RabbitMQ's messages* recipe, we have seen how to trace messages going to/from RabbitMQ.

However, it is not always possible, or desirable, to stop a running client (or a RabbitMQ server), modify its connection port, and point it to a different one; we just want to monitor the messages that are passing in real-time, impacting the system activity as little as possible.

[However, it's possible to activate the firehose tracer as seen in the recipe, *Tracing RabbitMQ's ongoing activity*.]

Wireshark is a free network analysis tool that has the capability to decode AMQP messages. This tool can be used either on the client side or on the server side to monitor the AMQP traffic flow seamlessly.

Getting ready

To exercise this recipe, you need RabbitMQ up and running and the RabbitMQ Java client library.

How to do it...

In the following steps, we are going to see how to use Wireshark to trace the AMQP messages:

1. If not already available on your system, download and install Wireshark from `http://www.wireshark.org/`. You can also install it for your distribution if it is available, for example, with:

   `yum install wireshark-gnome`

2. Start Wireshark on Linux from the `root` user.

3. Start to capture from the loopback interface.

4. From a terminal from the Java client library path, run the command:

   `./runjava.sh com.rabbitmq.examples.PerfTest -C 1 -D 1`

5. Stop the acquisition from the Wireshark GUI and analyze the captured AMQP traffic.

How it works...

Using Wireshark, it is possible to inspect the AMQP traffic exiting or entering a server that is hosting a RabbitMQ server or client.

In our example, we have captured the network traffic running both the client and the server on the same machine, thus connecting in `localhost`. That's why we were capturing the traffic from the loopback interface (step 3 of the previous steps).

Otherwise, we should capture the traffic from the network interface, usually eth0 or something similar.

> While on Linux, it's possible to capture traffic directed to `localhost`; the same does not apply to Windows. In this case, the client and the server must be on two different machines, and the capture must be activated on the network interface (either physical or virtual), thus connecting them.

So, in order to run the Wireshark graphical user interface, in case the RabbitMQ client and the server run on the same node, you need to select the loopback interface, as shown in the following screenshot:

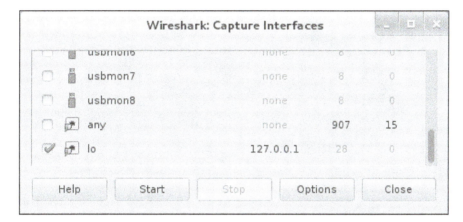

> On Linux, when you install the Wireshark package, you usually will have the command line interface only, **tshark**. To have Wireshark with the GUI installed, you have to install the appropriate package. For example, on Fedora, you have to install the `wireshark-gnome` package.

Once the AMQP traffic has travelled through the loopback interface, it has been captured by Wireshark.

The experiment run in step 4 of the previous steps actually starts both a producer and a consumer with two separated connections.

In order to highlight it, find a packet described as `Basic.PublishContent-Header`, right-click on it, and select `Follow TCP stream`. You can then close the window showing the payload dialogue between the client and the server. In the main window, you can now see the network packets that are exchanged between the client and the server, as shown in the following screenshot:

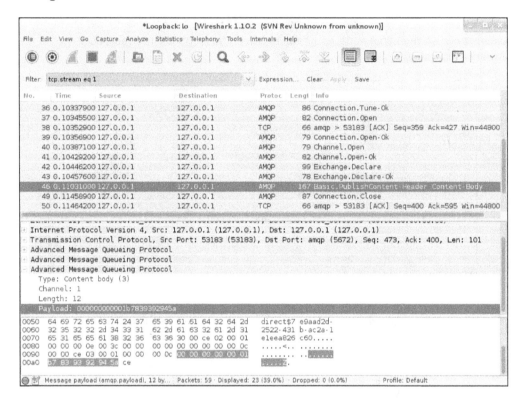

In the same way, you can select the traffic exiting the RabbitMQ server, as shown in the following screenshot:

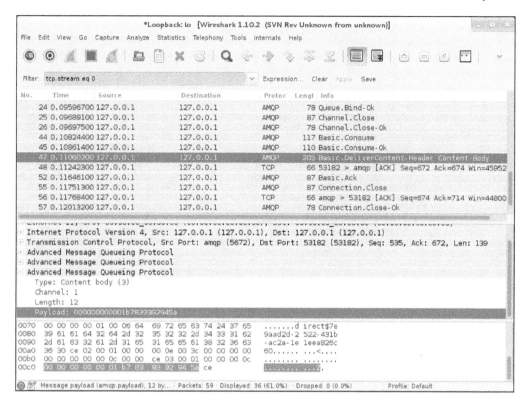

In the previous two screenshots, we have highlighted the AMQP payload of both messages, but you will find plenty of details in the AMQP traffic, thanks to the fact that Wireshark includes a very complete AMQP dissector.

There's more...

In case RabbitMQ is configured to use SSL and you want to analyze the encrypted traffic, this is possible under some given conditions by properly configuring the SSL public/private keys in the Wireshark configuration.

Find more information at `http://wiki.wireshark.org/SSL`.

See also

You can find some references to the Wireshark AMQP dissector at `http://wiki.wireshark.org/AMQP`.

Index

H

ha-mirror plugin 157
ha-mode: all 148
handleDelivery() callback 180
ha-policies 147
HAProxy
  URL 141
high-availability
  technologies, combining 167-169
High Performance Erlang (HiPE) 187

I

IBM MQ-Series 5
iPhone to RabbitMQ
  app, binding with MQTT 88-92

J

Jackson library
  URL 246
Java AMQP client library
  URL 6
JSON
  about 112
  body serialization, using with 17-19
  URL 112

L

Little Lemur plan 240
load balancer
  about 138
  introducing, to consumers 139-141
load spikes
  dealing with 231-236
localhost cluster
  about 130
  creating 130-132
londonorders queue 165

M

main() method 175
master image
  creating 219, 220

Mercurial
  URL 205
message 32
message destinations
  embedding, within messages 57, 58
message processing
  guaranteeing 32-34
message properties
  about 36
  accessing 36-38
Message Queue Telemetry Transport. *See* MQTT
message routing
  working with 28-32
messages
  between couple of brokers, distributing 157-160
  broadcasting 24-27
  consuming 14, 15
  distributing, to consumers 34, 35
  expiring 44, 45
  expiring, on specific queues 45-47
  forwarding 163-167
  message destinations, embedding within 57, 58
  producing 10-12
  publishing, from Android 98-100
messages, RabbitMQ
  debugging 259, 260
messaging
  about 181
  RPC, using with 20-22
  used, for updating Google Maps on Android 93-97
Microsoft MSMQ 5
Microsoft Windows Presentation Foundation (WPF) 86
mirror-all parameter 148
mirroring
  URL 157
mirror policies
  optimizing 154-156
MITM (man-in-the-middle) attacks 69
Mnesia database directory 261

Thank you for buying
RabbitMQ Cookbook

About Packt Publishing

Packt, pronounced 'packed', published its first book "*Mastering phpMyAdmin for Effective MySQL Management*" in April 2004 and subsequently continued to specialize in publishing highly focused books on specific technologies and solutions.

Our books and publications share the experiences of your fellow IT professionals in adapting and customizing today's systems, applications, and frameworks. Our solution based books give you the knowledge and power to customize the software and technologies you're using to get the job done. Packt books are more specific and less general than the IT books you have seen in the past. Our unique business model allows us to bring you more focused information, giving you more of what you need to know, and less of what you don't.

Packt is a modern, yet unique publishing company, which focuses on producing quality, cutting-edge books for communities of developers, administrators, and newbies alike. For more information, please visit our website: www.packtpub.com.

About Packt Open Source

In 2010, Packt launched two new brands, Packt Open Source and Packt Enterprise, in order to continue its focus on specialization. This book is part of the Packt Open Source brand, home to books published on software built around Open Source licences, and offering information to anybody from advanced developers to budding web designers. The Open Source brand also runs Packt's Open Source Royalty Scheme, by which Packt gives a royalty to each Open Source project about whose software a book is sold.

Writing for Packt

We welcome all inquiries from people who are interested in authoring. Book proposals should be sent to author@packtpub.com. If your book idea is still at an early stage and you would like to discuss it first before writing a formal book proposal, contact us; one of our commissioning editors will get in touch with you.

We're not just looking for published authors; if you have strong technical skills but no writing experience, our experienced editors can help you develop a writing career, or simply get some additional reward for your expertise.

open source*
community experience distilled

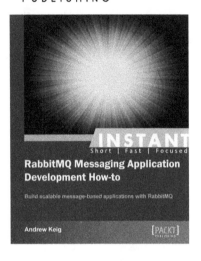

Instant RabbitMQ Messaging Application Development How-to

ISBN: 978-1-78216-574-3   Paperback: 54 pages

Build scalable message-based applications with RabbitMQ

1. Learn something new in an Instant! A short, fast, focused guide delivering immediate results

2. Learn how to build message-based applications with RabbitMQ using a practical Node.js e-commerce example

3. Implement various messaging patterns including asynchronous work queues, publish subscribe and topics

Instant Apache Camel Messaging System

ISBN: 978-1-78216-534-7   Paperback: 78 pages

Tackle integration problems and learn practical ways to make data flow between your application and other systems using Apache Camel

1. Learn something new in an Instant! A short, fast, focused guide delivering immediate results

2. Use Apache Camel to connect your application to different systems

3. Test your Camel application using unit tests, mocking, and component substitution

4. Configure your Apache Camel application using the Spring Framework

Please check **www.PacktPub.com** for information on our titles

Instant Apache Camel Message Routing

ISBN: 978-1-78328-347-7   Paperback: 62 pages

Route, transform, spilt, multicast messages, and do much more with Camel

1. Learn something new in an Instant! A short, fast, focused guide delivering immediate results

2. Learn how to use Enterprise Integration Patterns for message routing

3. Learn how Camel works and how it integrates disparate systems

4. Learn how to test and monitor Camel applications

Instant Spring for Android Starter

ISBN: 978-1-78216-190-5   Paperback: 72 pages

Leverage Spring for Android to create RESTful and OAuth Android apps

1. Learn something new in an Instant! A short, fast, focused guide delivering immediate results

2. Learn what Spring for Android adds to the Android developer toolkit

3. Learn how to debug your Android communication layer observing HTTP requests and responses

4. Use OAuth to authenticate your users and your app against popular service providers (Google, Facebook, Twitter, and so on)

Please check **www.PacktPub.com** for information on our titles